BRITIS  No.3

MU ニ UNITS
& ON-TRACK
MACHINES

TWENTY-FOURTH EDITION
2011

The complete guide to all Diesel Multiple
Units and On-Track Machines which operate
on the national railway network

Robert Pritchard, Peter Fox & Peter Hall

ISBN 978 1902 336 81 7

© 2010. Platform 5 Publishing Ltd., 3 Wyvern House, Sark Road, Sheffield,
S2 4HG, England.

Printed in England by Information Press, Eynsham, Oxford.

CONTENTS

PROVISION OF INFORMATION

This book has been compiled with care to be as accurate as possible, but in some cases information is not officially available and the publisher cannot be held responsible for any errors or omissions. We would like to thank the companies and individuals which have been co-operative in supplying information to us. The authors of this series of books are always pleased to receive notification from readers of any inaccuracies readers may find in the series, to enhance future editions. Please send comments to:

Robert Pritchard, Platform 5 Publishing Ltd., 3 Wyvern House, Sark Road, Sheffield, S2 4HG, England.

e-mail: robert@platform5.com **Tel:** 0114 255 2625 **Fax:** 0114 255 2471

This book is updated to information received by 4 October 2010.

UPDATES

This book is updated to the Stock Changes given in **Today's Railways UK 10** (November 2010). Readers are therefore advised to update this book from the official Platform 5 Stock Changes published every month in **Today's Railways UK** magazine, starting with issue 108.

The Platform 5 magazine **Today's Railways UK** contains news and rolling stock information on the railways of Britain and Ireland and is published on the second Monday of every month.

Front cover photograph: One of the five Grand Central 180s, 180 112 "JAMES HERRIOT" passes Sharlston, between Wakefield and Streethouse, with the 10.2 Bradford Interchange–London King's Cross on 16/06/10. **Ross Byers**

BRITAIN'S RAILWAY SYSTEM

INFRASTRUCTURE & OPERATION

Britain's national railway infrastructure is owned by a "not for dividend" company, Network Rail. Many stations and maintenance depots are leased to and operated by Train Operating Companies (TOCs), but some larger stations remain under Network Rail control. The only exception is the infrastructure on the Isle of Wight, which is nationally owned and is leased to South West Trains.

Trains are operated by TOCs over Network Rail, regulated by access agreements between the parties involved. In general, TOCs are responsible for the provision and maintenance of the locos, rolling stock and staff necessary for the direct operation of services, whilst NR is responsible for the provision and maintenance of the infrastructure and also for staff to regulate the operation of services.

DOMESTIC PASSENGER TRAIN OPERATORS

The large majority of passenger trains are operated by the TOCs on fixed term franchises. Franchise expiry dates are shown in the list of franchisees below:

Franchise	Franchisee	Trading Name
Chiltern Railways	Deutsche Bahn (until 31 December 2021)	Chiltern Railways
Cross-Country[1]	Deutsche Bahn (Arriva) (until 1 November 2013)	CrossCountry
East Midlands[2]	Stagecoach Holdings plc (until 11 November 2013)	East Midlands Trains
Greater Western[3]	First Group plc (until 1 April 2013)	First Great Western
Greater Anglia	National Express Group plc (until 14 October 2011)	National Express East Anglia
Integrated Kent[4]	GoVia Ltd. (Go-Ahead/Keolis) (until 3 March 2012)	Southeastern
InterCity East Coast[5]		East Coast
InterCity West Coast	Virgin Rail Group Ltd. (until 31 March 2012)	Virgin Trains
London Rail[6]	MTR/Deutsche Bahn (until 14 March 2014)	London Overground
LTS Rail[7]	National Express Group plc (until further notice)	c2c
Merseyrail Electrics[8]	Serco/NedRail (until 20 July 2028)	Merseyrail
Northern Rail	Serco/Abellio (until 15 September 2013)	Northern
ScotRail	First Group plc (until 8 November 2014)	ScotRail
South Central[9]	GoVia Ltd. (Go-Ahead/Keolis) (until 25 July 2015)	Southern

South Western[10]	Stagecoach Holdings plc (until 4 February 2014)	South West Trains
Thameslink/Great Northern[11]	First Group plc (until 1 April 2012)	First Capital Connect
Trans-Pennine Express[12]	First Group/Keolis (until 1 February 2012)	TransPennine Express
Wales & Borders	Deutsche Bahn (Arriva) (until 6 December 2018)	Arriva Trains Wales
West Midlands[13]	GoVia Ltd. (Go-Ahead/Keolis) (until 19 September 2013)	London Midland

Notes:

[1] Awarded for six years to 2013 with an extension for a further two years and five months to 1 April 2016 if performance targets are met.

[2] Awarded for six years to 2013 with an extension for a further one year and five months to 1 April 2015 if performance targets are met.

[3] Awarded for seven years to 2013 with an extension for a further three years to 1 April 2016 if performance targets are met.

[4] The Integrated Kent franchise started on 1 April 2006 for an initial period of six years to 2012, with an extension for a further two years to 1 April 2014 if performance targets are met.

[5] Currently run on an interim basis by DfT management company Directly Operated Railways (trading as East Coast) following financial difficulties experienced by National Express Group.

[6] The London Rail Concession is different from all other rail franchises, as fares and service levels are set by Transport for London instead of the DfT. Incorporates the North and West London lines, the Gospel Oak–Barking line and Euston–Watford local services.

[7] Tendering for the new LTS Rail franchise, to be called Essex Thameside, has been delayed whilst the Government conducts a consultation exercise on the future of rail franchising policy (the franchise has been due to finish on 28 May 2011). The Greater Anglia franchise has also been given a short extension for the same reason.

[8] Now under control of Merseytravel PTE instead of the DfT. Franchise due to be reviewed after seven years (in July 2010) and then every five years to fit in with the Merseyside Local Transport Plan.

[9] Awarded for five years and ten months to 2015 with a possible extension for a further two years to 25 July 2017.

[10] Awarded for seven years to 2014 with an extension for a further three years to 4 February 2017 if performance targets are met.

[11] Awarded for six years to 2012 with an extension for up to a further three years to 1 April 2015 if performance targets are met.

[12] Awarded for eight years to 2012 with an extension a further five years to 1 February 2017 if performance targets are met.

[13] Awarded for six years to 2013 with an extension for a further two years to 19 September 2015 if performance targets are met.

All new franchises officially start at 02.00 on the first day. Because of this the finishing date of an old franchise and the start date of its successor are the same.

Where termination dates are dependent on performance targets being met, the earliest possible termination date is given. However, with Merseyrail the termination date is based on the maximum franchise length.

The following operators run non-franchised services only:

Operator	Trading Name	Route
BAA	Heathrow Express	London Paddington–Heathrow Airport
First Hull Trains	First Hull Trains	London King's Cross–Hull
Grand Central	Grand Central	London King's Cross–Sunderland/ Bradford Interchange
North Yorkshire Moors Railway Enterprises	North Yorkshire Moors Railway	Pickering–Grosmont–Whitby/ Battersby
West Coast Railway Company	West Coast Railway Company	Birmingham–Stratford-upon-Avon Fort William–Mallaig* York–Leeds–York–Scarborough* Machynlleth–Porthmadog/Pwllheli*
Wrexham, Shropshire & Marylebone Railway	Wrexham & Shropshire	London Marylebone–Wrexham General

* Special summer-dated services only.

INTERNATIONAL PASSENGER OPERATIONS

Eurostar (UK) operates passenger services between the UK and mainland Europe, jointly with the national operators of France (SNCF) and Belgium (SNCB/NMBS). Eurostar (UK) is a subsidiary of London & Continental Railways, which is jointly owned by National Express Group and British Airways.

In addition, a service for the conveyance of accompanied road vehicles through the Channel Tunnel is provided by the tunnel operating company, Eurotunnel.

FREIGHT TRAIN OPERATIONS

The following operators operate freight services or empty passenger stock workings under "Open Access" arrangements:

Colas Rail
DB Schenker Rail (UK)
Direct Rail Services (DRS)
Europorte2 (Eurotunnel)
Freightliner
GB Railfreight (owned by Eurotunnel)
West Coast Railway Company

INTRODUCTION

DMU CLASSES

DMU Classes are listed in class number order. Principal details and dimensions are quoted for each class in metric and/or imperial units as considered appropriate bearing in mind common usage in the UK.

All dimensions and weights are quoted for vehicles in an "as new" condition with all necessary supplies (e.g. oil, water, sand) on board. Dimensions are quoted in the order Length – Width. All lengths quoted are over buffers or couplers as appropriate. Where two lengths are quoted, the first refers to outer vehicles in a set and the second to inner vehicles. All width dimensions quoted are maxima.

NUMERICAL LISTINGS

DMUs are listed in numerical order of set – using current numbers as allocated by the RSL. Individual "loose" vehicles are listed in numerical order after vehicles formed into fixed formations. Where sets or vehicles have been renumbered in recent years, former numbering detail is shown in parentheses. Each entry is laid out as in the following example:

RSL Set No.	Detail	Livery	Owner	Operator	Depot	Formation		Name
142 073	v	**AV**	A	*AW*	CF	55723	55769	Myfanwy

Detail Differences. Detail differences which currently affect the areas and types of train which vehicles may work are shown, plus differences in interior layout. Where such differences occur within a class, these are shown either in the heading information or alongside the individual set or vehicle number. The following standard abbreviations are used:

e European Railway Traffic Management System (ERTMS) signalling equipment fitted

r Radio Electronic Token Block (RETB) equipment.

Use of the above abbreviations indicates the equipment indicated is normally operable. Meaning of non-standard abbreviations is detailed in individual class headings.

Set Formations. Regular set formations are shown where these are normally maintained. Readers should note set formations might be temporarily varied from time to time to suit maintenance and/or operational requirements. Vehicles shown as "spare" are not formed in any regular set formation.

Codes. Codes are used to denote the livery, owner, operation and depot of each unit. Details of these will be found in section 6 of this book. Where a unit or spare car is off-lease, the operator column will be left blank.

Names. Only names carried with official sanction are listed. As far as possible names are shown in UPPER/lower case characters as actually shown on the name carried on the vehicle(s). Unless otherwise shown, complete units are regarded as named rather than just the individual car(s) which carry the name

GENERAL INFORMATION

CLASSIFICATION AND NUMBERING

First generation ("Heritage") DMUs are classified in the series 100–139.
Second generation DMUs are classified in the series 140–199.
Diesel-electric multiple units are classified in the series 200–249.
Service units are classified in the series 930–999.
First and second generation individual cars are numbered in the series 50000–59999 and 79000–79999.

DEMU individual cars are numbered in the series 60000–60999, except for a few former EMU vehicles which retain their EMU numbers.

Service stock individual cars are numbered in the series 975000–975999 and 977000–977999, although this series is not exclusively used for DMU vehicles.

OPERATING CODES

These codes are used by train operating company staff to describe the various different types of vehicles and normally appear on data panels on the inner (i.e. non driving) ends of vehicles.

The first part of the code describes whether or not the car has a motor or a driving cab as follows:

DM Driving motor.
M Motor
DT Driving trailer
T Trailer

The next letter is a "B" for cars with a brake compartment.

This is followed by the saloon details:

F First
S Standard
C Composite
so denotes a semi-open vehicle (part compartments, part open). All other vehicles are assumed to consist solely of open saloons.

L denotes a vehicle with a toilet.

Finally vehicles with a buffet are suffixed RB or RMB for a miniature buffet.

Where two vehicles of the same type are formed within the same unit, the above codes may be suffixed by (A) and (B) to differentiate between the vehicles.

A composite is a vehicle containing both first and standard class accommodation, whilst a brake vehicle is a vehicle containing separate specific accommodation for the conductor.

Special Note: Where vehicles have been declassified, the correct operating code which describes the actual vehicle layout is quoted in this publication.

BUILD DETAILS

Lot Numbers

Vehicles ordered under the auspices of BR were allocated a Lot (batch) number when ordered and these are quoted in class headings and sub-headings.

ACCOMMODATION

The information given in class headings and sub-headings is in the form F/S nT (or TD) nW. For example 12/54 1T 1W denotes 12 first class and 54 standard class seats, one toilet and one space for a wheelchair. A number in brackets (i.e. (2)) denotes tip-up seats (in addition to the fixed seats). Tip-up seats in vestibules do not count. The seating layout of open saloons is shown as 2+1, 2+2 or 3+2 as the case may be. Where units have first class accommodation as well as standard and the layout is different for each class then these are shown separately prefixed by "1:" and "2:". TD denotes a toilet suitable for use by a disabled person.

ABBREVIATIONS

The following abbreviations are used in class headings and also throughout this publication:

BR	British Railways.
BSI	Bergische Stahl Industrie.
DEMU	Diesel Electric Multiple Unit.
DMU	Diesel Multiple Unit (general term).
EMU	Electric Multiple Unit.
kN	kilonewtons.
km/h	kilometres per hour.
kW	kilowatts.
LT	London Transport.
LUL	London Underground Limited.
m.	metres.
m.p.h.	miles per hour.
t.	tonnes.

ON-TRACK MACHINES

From this edition we are pleased to be able to include On-Track Machines, such as Tampers, Stoneblowers, Multi Purpose Vehicles and Snowploughs. Please note that vehicles that can only operate within engineering possessions referred to as "On-Track Plant" (such as Road Rail Vehicles) are not listed.

1. DIESEL MECHANICAL & DIESEL HYDRAULIC UNITS

FIRST GENERATION UNITS

CLASS 121   PRESSED STEEL SUBURBAN

First generation units used by Chiltern Railways on selected Aylesbury–Princes Risborough services (121 020) and by Arriva Trains Wales on Cardiff Queen Street–Cardiff Bay shuttles (121 032).

Construction: Steel.
Engines: Two Leyland 1595 of 112 kW (150 h.p.) at 1800 r.p.m.
Transmission: Mechanical. Cardan shaft and freewheel to a four-speed epicyclic gearbox and final drive.
Bogies: DD10.
Brakes: Vacuum.
Couplers: Screw.
Dimensions: 20.45 x 2.82 m.
Gangways: Non gangwayed single cars with cabs at each end.
Wheel arrangement: 1-A + A-1.
Doors: Manually-operated slam.
Maximum Speed: 70 m.p.h.
Seating Layout: 3+2 facing.
Multiple Working: "Blue Square" coupling code. First Generation vehicles cannot be coupled to Second Generation units.

55020/55032. DMBS. Lot No. 30518 1960/1961. –/65. 38.0 t.

Non-standard livery: 121 020 All over Chiltern blue with a silver stripe.

Notes: Fitted with central door locking.

121 020 formerly in departmental use as unit 960 002 (977722).

121 032 formerly in departmental use as 977842, and more recently in preservation at The Railway Age, Crewe.

| 121 020 | **0** | CR | *CR* | AL | 55020 |
| 121 032 | **AV** | AW | *AW* | CF | 55032 |

PARRY PEOPLE MOVERS

CLASS 139   PPM-60

Gas/flywheel hybrid drive Railcars used on the Stourbridge Junction–Stourbridge Town branch.
Body construction: Stainless steel framework.
Chassis construction: Welded mild steel box section.
Primary Drive: Ford MVH420 2.3 litre 64 kW (86 h.p.) LPG fuel engine driving through Newage marine gearbox, Tandler bevel box and 4 "V" belt driver to flywheel.
Flywheel Energy Store: 500 kg, 1 m diameter, normal operational speed range 1000–1500 r.p.m.
Final transmission: 4 "V" belt driver from flywheel to Tandler bevel box, Linde hydrostatic transmission and spiral bevel gearbox at No. 2 end axle.
Braking: Normal service braking by regeneration to flywheel (1 m/s/s); emergency/parking braking by sprung-on, air-off disc brakes (3 m/s/s).
Maximum Speed: 45 m.p.h.
Dimensions: 8.7 x 2.4 m.
Doors: Deans powered doors, double-leaf folding (one per side).
Seating Layout: 1+1 unidirectional/facing.
Multiple Working: Not applicable.

39001–39002. DMS. Main Road Sheet Metal, Leyland 2007–08. –/20 1W. 12.5 t.

| 139 001 | **LM** | P | *LM* | SJ | 39001 |
| 139 002 | **LM** | P | *LM* | SJ | 39002 |

SECOND GENERATION UNITS

All units in this section have air brakes and are equipped with public address, with transmission equipment on driving vehicles and flexible diaphragm gangways. Except where otherwise stated, transmission is Voith 211r hydraulic with a cardan shaft to a Gmeinder GM190 final drive.

CLASS 142   PACER   BREL DERBY/LEYLAND

DMS–DMSL.

Construction: Steel underframe, aluminium alloy body and roof. Built from Leyland National bus parts on four-wheeled underframes.
Engines: One Cummins LT10-R of 165 kW (225 h.p.) at 1950 r.p.m.
Couplers: BSI at outer ends, bar within unit.
Dimensions: 15.55 x 2.80 m.
Gangways: Within unit only.   **Wheel Arrangement:** 1-A + A-1.
Doors: Twin-leaf inward pivoting.   **Maximum Speed:** 75 m.p.h.
Seating Layout: 3+2 mainly unidirectional bus/bench style unless stated.
Multiple Working: Within class and with Classes 143, 144, 150, 153, 155, 156, 158 and 159.

55542–55591. DMS. Lot No. 31003 1985–1986. –/62 (c –/46(6) 2W, s –/56, t –/53 or 55 1W, u –/52 or 54 1W). 24.5 t.
55592–55641. DMSL. Lot No. 31004 1985–1986. –/59 1T (c –/44(6) 1T 2W, s –/50 1T, u –/60 1T). 25.0 t.
55701–55746. DMS. Lot No. 31013 1986–1987. –/62 (c –/46(6) 2W, s –/56, t –/53 or 55 1W, u –/52 or 54 1W). 24.5 t.
55747–55792. DMSL. Lot No. 31014 1986–1987. –/59 1T (c –/44(6) 1T 2W, s –/50 1T, u –/60 1T). 25.0 t.

Notes:

c Refurbished Arriva Trains Wales units. Fitted with 2+2 individual Chapman seating.
s Fitted with 2+2 individual high-back seating.
t Former First North Western facelifted units – DMS fitted with a luggage/bicycle rack and wheelchair space.
u Merseytravel units – Fitted with 3+2 individual low-back seating.

The following units are on sub-lease from Northern to First Great Western:
142 001/009/029/030/063/064/068.

142 001	t	**NW**	A	*GW*	EX	55542 55592
142 002	c	**AV**	A	*AW*	CF	55543 55593
142 003		**NO**	A	*NO*	NH	55544 55594
142 004	t	**NO**	A	*NO*	NH	55545 55595
142 005	t	**NO**	A	*NO*	NH	55546 55596
142 006	c	**AV**	A	*AW*	CF	55547 55597
142 007	t	**NO**	A	*NO*	NH	55548 55598
142 009	t	**NW**	A	*GW*	EX	55550 55600
142 010	c	**AV**	A	*AW*	CF	55551 55601
142 011	t	**NO**	A	*NO*	NH	55552 55602
142 012	t	**NO**	A	*NO*	NH	55553 55603
142 013		**NO**	A	*NO*	NH	55554 55604
142 014	t	**NO**	A	*NO*	NH	55555 55605
142 015	s	**NO**	A	*NO*	HT	55556 55606
142 016	s	**NO**	A	*NO*	HT	55557 55607
142 017	s	**NO**	A	*NO*	HT	55558 55608
142 018	s	**NO**	A	*NO*	HT	55559 55609
142 019	s	**NO**	A	*NO*	HT	55560 55610
142 020	s	**NO**	A	*NO*	HT	55561 55611
142 021	s	**NO**	A	*NO*	HT	55562 55612
142 022	s	**NO**	A	*NO*	HT	55563 55613
142 023	t	**NO**	A	*NO*	HT	55564 55614
142 024	s	**NO**	A	*NO*	HT	55565 55615
142 025	s	**NO**	A	*NO*	HT	55566 55616
142 026	s	**NO**	A	*NO*	HT	55567 55617
142 027	t	**NO**	A	*NO*	HT	55568 55618
142 028	t	**NO**	A	*NO*	NH	55569 55619
142 029		**NW**	A	*GW*	EX	55570 55620
142 030		**NW**	A	*GW*	EX	55571 55621
142 031	t	**NO**	A	*NO*	NH	55572 55622
142 032	t	**NO**	A	*NO*	NH	55573 55623
142 033	t	**NO**	A	*NO*	NH	55574 55624

142 034	t	**NO**	A	*NO*	HT	55575	55625	
142 035	t	**NO**	A	*NO*	NH	55576	55626	
142 036	t	**NO**	A	*NO*	NH	55577	55627	
142 037	t	**NO**	A	*NO*	NH	55578	55628	
142 038	t	**NO**	A	*NO*	NH	55579	55629	
142 039	t	**NO**	A	*NO*	NH	55580	55630	
142 040	t	**NO**	A	*NO*	NH	55581	55631	
142 041	u	**NO**	A	*NO*	NH	55582	55632	
142 042	u	**NO**	A	*NO*	NH	55583	55633	
142 043	u	**NO**	A	*NO*	NH	55584	55634	
142 044	u	**NO**	A	*NO*	NH	55585	55635	
142 045	u	**NO**	A	*NO*	NH	55586	55636	
142 046	u	**NO**	A	*NO*	NH	55587	55637	
142 047	u	**NO**	A	*NO*	NH	55588	55638	
142 048	u	**NO**	A	*NO*	NH	55589	55639	
142 049	u	**NO**	A	*NO*	NH	55590	55640	
142 050	s	**NO**	A	*NO*	HT	55591	55641	
142 051	u	**NO**	A	*NO*	NH	55701	55747	
142 052	u	**NO**	A	*NO*	NH	55702	55748	
142 053	u	**NO**	A	*NO*	NH	55703	55749	
142 054	u	**NO**	A	*NO*	NH	55704	55750	
142 055	u	**NO**	A	*NO*	NH	55705	55751	
142 056	u	**NO**	A	*NO*	NH	55706	55752	
142 057	u	**NO**	A	*NO*	NH	55707	55753	
142 058	u	**NO**	A	*NO*	NH	55708	55754	
142 060	t	**NO**	A	*NO*	NH	55710	55756	
142 061	t	**NO**	A	*NO*	NH	55711	55757	
142 062	t	**NO**	A	*NO*	NH	55712	55758	
142 063	t	**NW**	A	*GW*	EX	55713	55759	
142 064	t	**NW**	A	*GW*	EX	55714	55760	
142 065	s	**NO**	A	*NO*	HT	55715	55761	
142 066	s	**NO**	A	*NO*	HT	55716	55762	
142 067		**NO**	A	*NO*	HT	55717	55763	
142 068	t	**NW**	A	*GW*	EX	55718	55764	
142 069	c	**AV**	A	*AW*	CF	55719	55765	
142 070	t	**NO**	A	*NO*	HT	55720	55766	
142 071	s	**NO**	A	*NO*	HT	55721	55767	
142 072	c	**AV**	A	*AW*	CF	55722	55768	
142 073	c	**AV**	A	*AW*	CF	55723	55769	Myfanwy
142 074	c	**AV**	A	*AW*	CF	55724	55770	
142 075	c	**AV**	A	*AW*	CF	55725	55771	
142 076	c	**AV**	A	*AW*	CF	55726	55772	
142 077	c	**AV**	A	*AW*	CF	55727	55773	
142 078	s	**NO**	A	*NO*	HT	55728	55774	
142 079	s	**NO**	A	*NO*	HT	55729	55775	
142 080	c	**AV**	A	*AW*	CF	55730	55776	
142 081	c	**AV**	A	*AW*	CF	55731	55777	
142 082	c	**AV**	A	*AW*	CF	55732	55778	
142 083	c	**AV**	A	*AW*	CF	55733	55779	
142 084	s	**NO**	A	*NO*	HT	55734	55780	
142 085	c	**AV**	A	*AW*	CF	55735	55781	

142 086	s	**NO**	A	*NO*	HT	55736 55782
142 087	s	**NO**	A	*NO*	HT	55737 55783
142 088	s	**NO**	A	*NO*	HT	55738 55784
142 089	s	**NO**	A	*NO*	HT	55739 55785
142 090	s	**NO**	A	*NO*	HT	55740 55786
142 091	s	**NO**	A	*NO*	HT	55741 55787
142 092	s	**NO**	A	*NO*	HT	55742 55788
142 093	s	**NO**	A	*NO*	HT	55743 55789
142 094	s	**NO**	A	*NO*	HT	55744 55790
142 095	s	**NO**	A	*NO*	HT	55745 55791
142 096	s	**NO**	A	*NO*	HT	55746 55792

CLASS 143   PACER   ALEXANDER/BARCLAY

DMS–DMSL. Similar design to Class 142, but bodies built by W. Alexander with Barclay underframes.

Construction: Steel underframe, aluminium alloy body and roof. Alexander bus bodywork on four-wheeled underframes.
Engines: One Cummins LT10-R of 165 kW (225 h.p.) at 1950 r.p.m.
Couplers: BSI at outer ends, bar within unit.
Dimensions: 15.45 x 2.80 m.
Gangways: Within unit only.   **Wheel Arrangement:** 1-A + A-1.
Doors: Twin-leaf inward pivoting.   **Maximum Speed:** 75 m.p.h.
Seating Layout: 2+2 high-back Chapman seating, mainly unidirectional.
Multiple Working: Within class and with Classes 142, 144, 150, 153, 155, 156, 158 and 159.

DMS. Lot No. 31005 Andrew Barclay 1985–1986. –/48(6) 2W. 24.0 t.
DMSL. Lot No. 31006 Andrew Barclay 1985–1986. –/44(6) 1T 2W. 24.5 t.

143 601	**AV**	BC	*AW*	CF	55642	55667	
143 602	**AV**	P	*AW*	CF	55651	55668	
143 603	**FI**	P	*GW*	EX	55658	55669	
143 604	**AV**	P	*AW*	CF	55645	55670	
143 605	**AV**	P	*AW*	CF	55646	55671	
143 606	**AV**	P	*AW*	CF	55647	55672	
143 607	**AV**	P	*AW*	CF	55648	55673	
143 608	**AV**	P	*AW*	CF	55649	55674	
143 609	**AV**	CC	*AW*	CF	55650	55675	Sir Tom Jones
143 610	**AV**	BC	*AW*	CF	55643	55676	
143 611	**FI**	P	*GW*	EX	55652	55677	
143 612	**FI**	P	*GW*	EX	55653	55678	
143 614	**AV**	BC	*AW*	CF	55655	55680	
143 616	**AV**	P	*AW*	CF	55657	55682	
143 617	**FI**	RI	*GW*	EX	55644	55683	
143 618	**FI**	RI	*GW*	EX	55659	55684	
143 619	**FI**	RI	*GW*	EX	55660	55685	
143 620	**FI**	P	*GW*	EX	55661	55686	
143 621	**FI**	P	*GW*	EX	55662	55687	
143 622	**AV**	P	*AW*	CF	55663	55688	
143 623	**AV**	P	*AW*	CF	55664	55689	

| 143 624 | **AV** | P | *AW* | CF | 55665 | 55690 |
| 143 625 | **AV** | P | *AW* | CF | 55666 | 55691 |

CLASS 144   PACER   ALEXANDER/BREL DERBY

DMS–DMSL or DMS–MS–DMSL. As Class 143, but underframes built by BREL.

Construction: Steel underframe, aluminium alloy body and roof. Alexander bus bodywork on four-wheeled underframes.
Engines: One Cummins LT10-R of 165 kW (225 h.p.) at 1950 r.p.m.
Couplers: BSI at outer ends, bar within unit.
Dimensions: 15.45/15.43 x 2.80 m.
Gangways: Within unit only.   **Wheel Arrangement:** 1-A + A-1.
Doors: Twin-leaf inward pivoting.   **Maximum Speed:** 75 m.p.h.
Seating Layout: 2+2 high-back Richmond seating, mainly unidirectional.
Multiple Working: Within class and with Classes 142, 143, 150, 153, 155, 156, 158 and 159.

DMS. Lot No. 31015 BREL Derby 1986–1987. –/45(3) 1W 24.0 t.
MS. Lot No. BREL Derby 31037 1987. –/58. 23.5 t.
DMSL. Lot No. BREL Derby 31016 1986–1987. –/42(3) 1T. 24.5 t.

Note: The centre cars of the 3-car units are owned by West Yorkshire PTE, although managed by Porterbrook Leasing Company.

144 001	**NO**	P	*NO*	NL	55801		55824
144 002	**NO**	P	*NO*	NL	55802		55825
144 003	**NO**	P	*NO*	NL	55803		55826
144 004	**NO**	P	*NO*	NL	55804		55827
144 005	**NO**	P	*NO*	NL	55805		55828
144 006	**NO**	P	*NO*	NL	55806		55829
144 007	**NO**	P	*NO*	NL	55807		55830
144 008	**NO**	P	*NO*	NL	55808		55831
144 009	**NO**	P	*NO*	NL	55809		55832
144 010	**NO**	P	*NO*	NL	55810		55833
144 011	**NO**	P	*NO*	NL	55811		55834
144 012	**NO**	P	*NO*	NL	55812		55835
144 013	**NO**	P	*NO*	NL	55813		55836
144 014	**NO**	P	*NO*	NL	55814	55850	55837
144 015	**NO**	P	*NO*	NL	55815	55851	55838
144 016	**NO**	P	*NO*	NL	55816	55852	55839
144 017	**NO**	P	*NO*	NL	55817	55853	55840
144 018	**NO**	P	*NO*	NL	55818	55854	55841
144 019	**NO**	P	*NO*	NL	55819	55855	55842
144 020	**NO**	P	*NO*	NL	55820	55856	55843
144 021	**NO**	P	*NO*	NL	55821	55857	55844
144 022	**NO**	P	*NO*	NL	55822	55858	55845
144 023	**NO**	P	*NO*	NL	55823	55859	55846

Name: 144 001 THE PENISTONE LINE PARTNERSHIP

CLASS 150/0   SPRINTER   BREL YORK

DMSL–MS–DMS. Prototype Sprinter.

Construction: Steel.
Engines: One Cummins NT-855-R4 of 213 kW (285 h.p.) at 2100 r.p.m.
Bogies: BX8P (powered), BX8T (non-powered).
Couplers: BSI at outer end of driving vehicles, bar non-driving ends.
Dimensions: 20.06/20.18 x 2.82 m.
Gangways: Within unit only.   **Wheel Arrangement:** 2-B + 2-B + B-2.
Doors: Twin-leaf sliding.   **Maximum Speed:** 75 m.p.h.
Seating Layout: 3+2 (mainly unidirectional).
Multiple Working: Within class and with Classes 142, 143, 144, 153, 155, 156, 158, 159, 170 and 172.

DMSL. Lot No. 30984 1984. –/72 1T. 35.4 t.
MS. Lot No. 30986 1984. –/92. 34.1 t.
DMS. Lot No. 30985 1984. –/76. 29.5 t.

150 001		CI	A	LM	TS	55200	55400	55300
150 002		CI	A	LM	TS	55201	55401	55301

CLASS 150/1   SPRINTER   BREL YORK

DMSL–DMS or DMSL–DMSL–DMS or DMSL–DMS–DMS.

Construction: Steel.
Engines: One Cummins NT855R5 of 213 kW (285 h.p.) at 2100 r.p.m.
Bogies: BP38 (powered), BT38 (non-powered).
Couplers: BSI.
Dimensions: 19.74 x 2.82 m.
Gangways: Within unit only.   **Wheel Arrangement:** 2-B (+ 2–B) + B-2.
Doors: Twin-leaf sliding.   **Maximum Speed:** 75 m.p.h.
Seating Layout: 3+2 facing as built but Centro units were reseated with mainly unidirectional seating.
Multiple Working: Within class and with Classes 142, 143, 144, 153, 155, 156, 158, 159, 170 and 172.

DMSL. Lot No. 31011 1985–1986. –/72 1T (c –/59 1TD (except 52144 which is –/62 1TD), t –/71 1T, u –/71 1T). 38.3 t.
DMS. Lot No. 31012 1985–1986. –/76 (c –/65, t –/73, u –/70). 38.1 t.

Notes: The centre cars of 3-car units are Class 150/2 vehicles. For details see Class 150/2.

Many London Overground and London Midland 150s will be reallocated elsewhere once new 172s are in traffic. 150 101/102/104/106/108/120/122/123/124/125/126/128/129/130/131 will transfer to First Great Western and Northern will also receive eight 2-car 150s.

c 3+2 Chapman seating.

150 003	u	WM	A	LM	TS	52103	57210	57103
150 005	u	CI	A	LM	TS	52105	52210	57105

150 007	u	CI	A	*LM*	TS	52107	52202	57107
150 009	u	CI	A	*LM*	TS	52109	57202	57109
150 010	u	WM	A	*LM*	TS	52110	57226	57110
150 011	u	CI	A	*LM*	TS	52111	52204	57111
150 012	u	CI	A	*LM*	TS	52112	57206	57112
150 013	u	CI	A	*LM*	TS	52113	52226	57113
150 014	u	CI	A	*LM*	TS	52114	57204	57114
150 015	u	CI	A	*LM*	TS	52115	52206	57115
150 016	u	CI	A	*LM*	TS	52116	57212	57116
150 017	u	CI	A	*LM*	TS	52117	57209	57117
150 018	u	WM	A	*LM*	TS	52118	52220	57118
150 019	u	CI	A	*LM*	TS	52119	57220	57119
150 101	u	CI	A	*LM*	TS	52101	57101
150 102	u	CI	A	*LM*	TS	52102	57102
150 104	u	CI	A	*LM*	TS	52104	57104
150 106	u	CI	A	*LM*	TS	52106	57106
150 108	u	CI	A	*LM*	TS	52108	57108
150 120	t	SL	A	*LO*	WN	52120	57120
150 121	u	SL	A	*GW*	PM	52121	57121
150 122	u	CI	A	*LM*	TS	52122	57122
150 123	t	SL	A	*LO*	WN	52123	57123
150 124	u	CI	A	*LM*	TS	52124	57124
150 125	u	CI	A	*LM*	TS	52125	57125
150 126	u	WM	A	*LM*	TS	52126	57126
150 127	t	SL	A	*GW*	PM	52127	57127
150 128	t	SL	A	*LO*	WN	52128	57128
150 129	t	SL	A	*LO*	WN	52129	57129
150 130	t	SL	A	*LO*	WN	52130	57130
150 131	t	SL	A	*LO*	WN	52131	57131
150 132	u	WM	A	*LM*	TS	52132	57132
150 133	c	NO	A	*NO*	NH	52133	57133
150 134	c	NO	A	*NO*	NH	52134	57134
150 135	c	NO	A	*NO*	NH	52135	57135
150 136	c	NO	A	*NO*	NH	52136	57136
150 137	c	NO	A	*NO*	NH	52137	57137
150 138	c	NO	A	*NO*	NH	52138	57138
150 139	c	NO	A	*NO*	NH	52139	57139
150 140	c	NO	A	*NO*	NH	52140	57140
150 141	c	NO	A	*NO*	NH	52141	57141
150 142	c	NO	A	*NO*	NH	52142	57142
150 143	c	NO	A	*NO*	NH	52143	57143
150 144	c	NO	A	*NO*	NH	52144	57144
150 145	c	NO	A	*NO*	NH	52145	57145
150 146	c	NO	A	*NO*	NH	52146	57146
150 147	c	NO	A	*NO*	NH	52147	57147
150 148	c	NO	A	*NO*	NH	52148	57148
150 149	c	NO	A	*NO*	NH	52149	57149
150 150	c	NO	A	*NO*	NH	52150	57150

▲ Parry People Mover 139 002 arrives at Stourbridge Junction with the 08.44 from Stourbridge Town on 11/09/09. **Alisdair Anderson**

▼ The Arriva Trains Wales Class 121 "Bubble Car" is used on the short Cardiff Bay branch. On 09/08/10 it leaves Cardiff Queen Street with the 15.48 to Cardiff Bay. **Robert Pritchard**

▲ Northern-liveried 142 038 is seen at Sheffield after arrival with the 09.13 from Huddersfield on 21/06/10. **Robert Pritchard**

▼ Arriva Trains-liveried 143 625 arrives at Tenby with the 09.08 Carmarthen–Pembroke Dock on 20/06/10. **Mick Tindall**

▲ Northern-liveried 144 011 runs alongside the Sheffield & South Yorkshire Navigation (Rotherham Cut) at Rotherham Parkgate with the 18.26 Doncaster–Sheffield on 22/06/10. **Robert Pritchard**

▼ First Great Western local lines-liveried 150 219 arrives at Bradford-on-Avon with a Bristol–Westbury service on 01/05/10. **Mark Few**

▲ A pair of East Midlands Trains-liveried Class 153s, led by 153 383, leave Longport with the 17.48 Derby–Crewe on 15/05/10. **Cliff Beeton**

▼ Northern-liveried 155 343 leaves Cross Gates with the 08.48 Manchester Victoria–Selby on 24/04/10. **Robert Pritchard**

▲ Northern-liveried 156 448, carrying promotional vinyls for the Tyne Valley Line, is seen heading south at Parton on the Cumbrian Coast Line with the 09.40 Carlisle–Lancaster on 30/08/10. **Neil Gibson**

▲ The first Class 158 in the new ScotRail livery, 158 871, leaves Dunblane with the 13.58 to Edinburgh Waverley on 09/09/10. **Ian Lothian**

▼ South West Trains white-liveried 159 022 arrives at Salisbury with the 09.20 Exeter St Davids–London Waterloo on 30/03/09. **Andrew Mist**

▲ Chiltern Railways-liveried 165 028 pauses at Sudbury Hill Harrow with the 14.02 High Wycombe–London Marylebone on 08/05/08. **Robert Pritchard**

▼ First Great Western Dynamic Lines-liveried 166 219 arrives at Reading with the 13.21 London Paddington–Great Malvern on 04/06/10. **Jason Rogers**

▲ Chiltern Railways-liveried 168 214 passes Wormleighton Crossing, near Banbury, with the 12.52 Birmingham Snow Hill–London Marylebone on 01/09/10. **Paul Biggs**

▼ CrossCountry-liveried 170 114 passes Attenborough with the 07.49 Birmingham New Street–Nottingham on 21/04/10. **Robert Pritchard**

CLASS 150/2   SPRINTER   BREL YORK

DMSL–DMS.

Construction: Steel.
Engines: One Cummins NT855R5 of 213 kW (285 h.p.) at 2100 r.p.m.
Bogies: BP38 (powered), BT38 (non-powered).
Couplers: BSI.
Dimensions: 19.74 x 2.82 m.
Gangways: Throughout.   **Wheel Arrangement:** 2-B + B-2.
Doors: Twin-leaf sliding.   **Maximum Speed:** 75 m.p.h.
Seating Layout: 3+2 mainly unidirectional seating as built, but most units have now been refurbished with new 2+2 seating (see notes below).
Multiple Working: Within class and with Classes 142, 143, 144, 153, 155, 156, 158, 159, 170 and 172.

DMSL. Lot No. 31017 1986–1987. –/73 1T (c –/62 1TD, p –/60(4) 1T, u –/71 1T), v –/60(8) 1T, w –/60(8) 1T). 37.5 t.
DMS. Lot No. 31018 1986–1987. –/76 (c –/70, p –/56(10) 1W, u –/70), v –/56(15) 2W, w –/56(17) 2W, z –/68). 36.5 t.

Northern promotional vinyls: 150 228/268–271/273–277 Welcome to Yorkshire
150 272 R&B Festival week, Colne

Notes:

c   3+2 Chapman seating (former First North Western units).
p   Refurbished Arriva Trains Wales units with 2+2 Primarius seating.
v   Units refurbished for Valley Lines with 2+2 Chapman seating.
w   Units refurbished for First Great Western with 2+2 Chapman seating.

The following units are on sub-lease from Arriva Trains Wales to First Great Western: 150 267/278/279/281.

150 201	c	**NO**	A	*NO*	NH	52201	57201
150 203	c	**NO**	A	*NO*	NH	52203	57203
150 205	c	**NO**	A	*NO*	NH	52205	57205
150 207	c	**NO**	A	*NO*	NH	52207	57207
150 208	p	**AV**	P	*AW*	CF	52208	57208
150 211	c	**NO**	A	*NO*	NH	52211	57211
150 213	p	**AV**	P	*AW*	CF	52213	57213
150 214	u	**CI**	A	*LM*	TS	52214	57214
150 215	c	**NO**	A	*NO*	NH	52215	57215
150 216	u	**CI**	A	*LM*	TS	52216	57216
150 217	p	**AV**	P	*AW*	CF	52217	57217
150 218	c	**NO**	A	*NO*	NH	52218	57218
150 219	w	**FI**	P	*GW*	PM	52219	57219
150 221	w	**FI**	P	*GW*	PM	52221	57221
150 222	c	**NO**	A	*NO*	NH	52222	57222
150 223	c	**NO**	A	*NO*	NH	52223	57223
150 224	c	**NO**	A	*NO*	NH	52224	57224
150 225	c	**NO**	A	*NO*	NH	52225	57225
150 227	p	**AV**	P	*AW*	CF	52227	57227

150 228		NO	P	NO	NH	52228	57228
150 229	p	AV	P	AW	CF	52229	57229
150 230	w	AV	P	AW	CF	52230	57230
150 231	p	AV	P	AW	CF	52231	57231
150 232	w	FI	P	GW	PM	52232	57232
150 233	w	FI	P	GW	PM	52233	57233
150 234	w	FI	P	GW	PM	52234	57234
150 235	p	AV	P	AW	CF	52235	57235
150 236	w	AV	P	AW	CF	52236	57236
150 237	p	AV	P	AW	CF	52237	57237
150 238	w	FI	P	GW	PM	52238	57238
150 239	w	FI	P	GW	PM	52239	57239
150 240	w	AV	P	AW	CF	52240	57240
150 241	w	AV	P	AW	CF	52241	57241
150 242	w	AV	P	AW	CF	52242	57242
150 243	w	FI	P	GW	PM	52243	57243
150 244	w	FI	P	GW	PM	52244	57244
150 245	p	AV	P	AW	CF	52245	57245
150 246	w	FI	P	GW	PM	52246	57246
150 247	w	FI	P	GW	PM	52247	57247
150 248	w	FI	P	GW	PM	52248	57248
150 249	w	FI	P	GW	PM	52249	57249
150 250	p	AV	P	AW	CF	52250	57250
150 251	w	AV	P	AW	CF	52251	57251
150 252	p	AV	P	AW	CF	52252	57252
150 253	w	AV	P	AW	CF	52253	57253
150 254	w	AV	P	AW	CF	52254	57254
150 255	p	AV	P	AW	CF	52255	57255
150 256	w	AV	P	AW	CF	52256	57256
150 257	p	AV	P	AW	CF	52257	57257
150 258	p	AV	P	AW	CF	52258	57258
150 259	p	AV	P	AW	CF	52259	57259
150 260	p	AV	P	AW	CF	52260	57260
150 261	w	FI	P	GW	PM	52261	57261
150 262	p	AV	P	AW	CF	52262	57262
150 263	w	FI	P	GW	PM	52263	57263
150 264	p	AV	P	AW	CF	52264	57264
150 265	w	FI	P	GW	PM	52265	57265
150 266	w	FI	P	GW	PM	52266	57266
150 267	v	AV	P	GW	PM	52267	57267
150 268		NO	P	NO	NH	52268	57268
150 269		NO	P	NO	NH	52269	57269
150 270		NO	P	NO	NH	52270	57270
150 271		NO	P	NO	NH	52271	57271
150 272		NO	P	NO	NH	52272	57272
150 273		NO	P	NO	NH	52273	57273
150 274		NO	P	NO	NH	52274	57274
150 275		NO	P	NO	NH	52275	57275
150 276		NO	P	NO	NH	52276	57276
150 277		NO	P	NO	NH	52277	57277
150 278	v	AV	P	GW	PM	52278	57278

150 279	v	**AV**	P	*GW*	PM	52279	57279
150 280	v	**AV**	P	*AW*	CF	52280	57280
150 281	v	**AV**	P	*GW*	PM	52281	57281
150 282	v	**AV**	P	*AW*	CF	52282	57282
150 283	p	**AV**	P	*AW*	CF	52283	57283
150 284	p	**AV**	P	*AW*	CF	52284	57284
150 285	p	**AV**	P	*AW*	CF	52285	57285

CLASS 153   SUPER SPRINTER   LEYLAND BUS

DMSL. Converted by Hunslet-Barclay, Kilmarnock from Class 155 2-car units.

Construction: Steel underframe, aluminium alloy body and roof. Built from Leyland National bus parts on bogied underframes.
Engine: One Cummins NT855R5 of 213 kW (285 h.p.) at 2100 r.p.m.
Bogies: One P3-10 (powered) and one BT38 (non-powered).
Couplers: BSI.
Dimensions: 23.21 x 2.70 m.
Gangways: Throughout.   **Wheel Arrangement:** 2-B.
Doors: Single-leaf sliding plug.   **Maximum Speed:** 75 m.p.h.
Seating Layout: 2+2 facing/unidirectional.
Multiple Working: Within class and with Classes 142, 143, 144, 150, 155, 156, 158, 159, 170 and 172.

52301–52335. DMSL. Lot No. 31026 1987–1988. Converted under Lot No. 31115 1991–1992. –/72(3) 1T 1W. (* –/66(3) 1T 1W, s –/72 1T 1W, t –/72(2) 1T 1W). 41.2 t.
57301–57335. DMSL. Lot No. 31027 1987–1988. Converted under Lot No. 31115 1991–1992. –/72(3) 1T 1W. (* –/66(3) 1T 1W). 41.2 t.

Notes: Cars numbered in the 573xx series were renumbered by adding 50 to their original number so that the last two digits correspond with the set number.

*   Refurbished East Anglia area units with a bicycle rack.
c   Chapman seating.
d   Richmond seating.
Units not shown as c or d were reseated using original Class 158 seats.

153 301	d	**NO**	A	*NO*	NL	52301	
153 302		**EM**	A	*EM*	NM	52302	
153 303	c	**AV**	A	*AW*	CF	52303	
153 304	ds	**NO**	A	*NO*	NL	52304	
153 305	d	**FI**	A	*GW*	EX	52305	
153 306	cr	**1**	P	*EA*	NC	52306	
153 307	d	**NO**	A	*NO*	NL	52307	
153 308		**EM**	A	*EM*	NM	52308	
153 309	cr	**AR**	P	*EA*	NC	52309	GERARD FIENNES
153 310	c	**EM**	P	*EM*	NM	52310	
153 311	c*	**EM**	P	*EM*	NM	52311	
153 312	s	**AV**	A	*AW*	CF	52312	
153 313	cs	**EM**	P	*EM*	NM	52313	
153 314	cr	**1**	P	*EA*	NC	52314	
153 315	ds	**NO**	A	*NO*	NL	52315	

153 316	c	**NO**	P	*NO*	NL	52316
153 317	ds	**NO**	A	*NO*	NL	52317
153 318	d	**FI**	A	*GW*	EX	52318
153 319	d	**EM**	A	*EM*	NM	52319
153 320	c	**AV**	P	*AW*	CF	52320
153 321	c	**EM**	P	*EM*	NM	52321
153 322	cr	**AR**	P	*EA*	NC	52322	BENJAMIN BRITTEN
153 323	c	**AV**	P	*AW*	CF	52323
153 324	c	**NO**	P	*NO*	NL	52324
153 325	c	**LM**	P	*LM*	TS	52325
153 326	c*	**EM**	P	*EM*	NM	52326
153 327	c	**AV**	A	*AW*	CF	52327
153 328	ds	**NO**	A	*NO*	NL	52328
153 329	c	**FI**	P	*GW*	EX	52329
153 330	cs	**NO**	P	*NO*	NL	52330
153 331	d	**NO**	A	*NO*	NL	52331
153 332	c	**NO**	P	*NO*	NL	52332
153 333	cs	**LM**	P	*LM*	TS	52333
153 334	ct	**LM**	P	*LM*	TS	52334
153 335	cr	**AR**	P	*EA*	NC	52335	MICHAEL PALIN
153 351	d	**NO**	A	*NO*	NL	57351
153 352	ds	**NO**	A	*NO*	NL	57352
153 353	c	**AV**	A	*AW*	CF	57353
153 354	c	**LM**	P	*LM*	TS	57354
153 355		**EM**	A	*EM*	NM	57355
153 356	c	**LM**	P	*LM*	TS	57356
153 357	d	**EM**	A	*EM*	NM	57357
153 358	c	**NO**	P	*NO*	NL	57358
153 359	c	**NO**	P	*NO*	NL	57359
153 360	c	**NO**	P	*NO*	NL	57360
153 361	cs	**FI**	P	*GW*	EX	57361
153 362	c	**AV**	A	*AW*	CF	57362	Dylan Thomas 1914–1953
153 363	cs	**NO**	P	*NO*	NL	57363
153 364	c	**LM**	P	*LM*	TS	57364
153 365	c	**LM**	P	*LM*	TS	57365
153 366	c	**LM**	P	*LM*	TS	57366
153 367	cs	**AV**	P	*AW*	CF	57367
153 368	d	**FI**	A	*GW*	EX	57368
153 369	c	**FI**	P	*GW*	EX	57369
153 370	d	**FI**	A	*GW*	EX	57370
153 371	c	**LM**	P	*LM*	TS	57371
153 372	d	**FI**	A	*GW*	EX	57372
153 373	d	**FI**	A	*GW*	EX	57373
153 374		**EM**	A	*EM*	NM	57374
153 375	c	**LM**	P	*LM*	TS	57375
153 376	c	**CT**	P	*EM*	NM	57376
153 377	d	**FI**	A	*GW*	EX	57377
153 378	d	**NO**	A	*NO*	NL	57378
153 379	c	**CT**	P	*EM*	NM	57379
153 380	d	**FI**	A	*GW*	EX	57380
153 381	c	**EM**	P	*EM*	NM	57381

53 382	d	**FI**	A	*GW*	EX	57382
53 383	c	**EM**	P	*EM*	NM	57383
53 384	c	**CT**	P	*EM*	NM	57384
53 385	c	**EM**	P	*EM*	NM	57385

CLASS 155   SUPER SPRINTER   LEYLAND BUS

DMSL–DMS.

Construction: Steel underframe, aluminium alloy body and roof. Built from Leyland National bus parts on bogied underframes.
Engines: One Cummins NT855R5 of 213 kW (285 h.p.) at 2100 r.p.m.
Bogies: One P3-10 (powered) and one BT38 (non-powered).
Couplers: BSI.
Dimensions: 23.21 x 2.70 m.
Gangways: Throughout.   **Wheel Arrangement:** 2-B + B-2.
Doors: Single-leaf sliding plug.   **Maximum Speed:** 75 m.p.h.
Seating Layout: 2+2 facing/unidirectional Chapman seating.
Multiple Working: Within class and with Classes 142, 143, 144, 150, 153, 156, 158, 159, 170 and 172.

DMSL. Lot No. 31057 1988. –/76 1TD 1W. 39.0 t.
DMS. Lot No. 31058 1988. –/80. 38.6 t.

Northern promotional vinyls:

55 341–347 Leeds–Bradford–Manchester route (the "Calder Valley").

Note: These units are owned by West Yorkshire PTE, although managed by Porterbrook Leasing Company.

55 341	**NO**	P	*NO*	NL	52341	57341
55 342	**NO**	P	*NO*	NL	52342	57342
55 343	**NO**	P	*NO*	NL	52343	57343
55 344	**NO**	P	*NO*	NL	52344	57344
55 345	**NO**	P	*NO*	NL	52345	57345
55 346	**NO**	P	*NO*	NL	52346	57346
55 347	**NO**	P	*NO*	NL	52347	57347

CLASS 156   SUPER SPRINTER   METRO-CAMMELL

DMSL–DMS.

Construction: Steel.
Engines: One Cummins NT855R5 of 213 kW (285 h.p.) at 2100 r.p.m.
Bogies: One P3-10 (powered) and one BT38 (non-powered).
Couplers: BSI.   **Dimensions:** 23.03 x 2.73 m.
Gangways: Throughout.   **Wheel Arrangement:** 2-B + B-2.
Doors: Single-leaf sliding.   **Maximum Speed:** 75 m.p.h.
Seating Layout: 2+2 facing/unidirectional.
Multiple Working: Within class and with Classes 142, 143, 144, 150, 153, 155, 158, 159, 170 and 172.

DMSL. Lot No. 31028 1988–1989. –/74 (†* –/72, c, t –/70, u –/68) 1TD 1W. 38.6 t.
DMS. Lot No. 31029 1987–1989. –/76 (d –/78, † –/74, t, u –/72) 36.1 t.

Advertising livery: 156 402 Chapelfield Shopping Centre (white & blue).

Northern promotional vinyls:

156 448 Hadrians Wall Country (Newcastle–Carlisle line).
156 461 Ravenglass & Eskdale Railway.
156 469 Bishop Auckland branch.
156 484 Settle & Carlisle line.
156 490 National Railway Museum.

Notes:

c Chapman seating.
d Richmond seating.

156 401	c*	**EM**	P	*EM*	DY	52401 57401
156 402	cr	**AL**	P	*EA*	NC	52402 57402
156 403	c*	**EM**	P	*EM*	DY	52403 57403
156 404	c*	**EM**	P	*EM*	DY	52404 57404
156 405	c*	**EM**	P	*EM*	DY	52405 57405
156 406	c*	**EM**	P	*EM*	DY	52406 57406
156 407	cr	**1**	P	*EA*	NC	52407 57407
156 408	c*	**EM**	P	*EM*	DY	52408 57408
156 409	cr	**1**	P	*EA*	NC	52409 57409
156 410	c*	**EM**	P	*EM*	DY	52410 57410
156 411	c*	**EM**	P	*EM*	DY	52411 57411
156 412	cr	**CT**	P	*EA*	NC	52412 57412
156 413	c*	**EM**	P	*EM*	DY	52413 57413
156 414	c*	**EM**	P	*EM*	DY	52414 57414
156 415	c*	**EM**	P	*EM*	DY	52415 57415
156 416	cr	**1**	P	*EA*	NC	52416 57416
156 417	cr	**1**	P	*EA*	NC	52417 57417
156 418	cr	**CT**	P	*EA*	NC	52418 57418
156 419	cr	**NX**	P	*EA*	NC	52419 57419
156 420	c	**NO**	P	*NO*	NH	52420 57420
156 421	c	**NO**	P	*NO*	NH	52421 57421
156 422	cr	**1**	P	*EA*	NC	52422 57422
156 423	c	**NO**	P	*NO*	NH	52423 57423
156 424	c	**NO**	P	*NO*	NH	52424 57424
156 425	c	**NO**	P	*NO*	NH	52425 57425
156 426	c	**NO**	P	*NO*	NH	52426 57426
156 427	c	**NO**	P	*NO*	NH	52427 57427
156 428	c	**NO**	P	*NO*	NH	52428 57428
156 429	c	**NO**	P	*NO*	NH	52429 57429
156 430	t	**SR**	A	*SR*	CK	52430 57430
156 431	t	**SR**	A	*SR*	CK	52431 57431
156 432	t	**SR**	A	*SR*	CK	52432 57432
156 433	t	**SR**	A	*SR*	CK	52433 57433
156 434	t	**SR**	A	*SR*	CK	52434 57434
156 435	t	**SR**	A	*SR*	CK	52435 57435
156 436	†	**SR**	A	*SR*	CK	52436 57436
156 437	t	**SR**	A	*SR*	CK	52437 57437
156 438	d	**NO**	A	*NO*	HT	52438 57438

156 439	t	SR	A	SR	CK	52439	57439
156 440	c	NO	P	NO	NH	52440	57440
156 441	c	NO	P	NO	NH	52441	57441
156 442	t	SR	A	SR	CK	52442	57442
156 443	d	NO	A	NO	HT	52443	57443
156 444	d	NO	A	NO	HT	52444	57444
156 445	u	SR	A	SR	CK	52445	57445
156 446	rt	FS	A	SR	CK	52446	57446
156 447	ru	FS	A	SR	CK	52447	57447
156 448	d	NO	A	NO	HT	52448	57448
156 449	u	FS	A	SR	CK	52449	57449
156 450	ru	FS	A	SR	CK	52450	57450
156 451	d	NO	A	NO	HT	52451	57451
156 452	c	NO	P	NO	NH	52452	57452
156 453	ru	FS	A	SR	CK	52453	57453
156 454	d	NO	A	NO	HT	52454	57454
156 455	c	NO	P	NO	NH	52455	57455
156 456	rt	FS	A	SR	CK	52456	57456
156 457	rt	FS	A	SR	CK	52457	57457
156 458	rt	FS	A	SR	CK	52458	57458
156 459	c	NO	P	NO	NH	52459	57459
156 460	c	NO	P	NO	NH	52460	57460
156 461	c	NO	P	NO	NH	52461	57461
156 462		FS	A	SR	CK	52462	57462
156 463	d	NO	A	NO	HT	52463	57463
156 464	c	NO	P	NO	NH	52464	57464
156 465	ru	FS	A	SR	CK	52465	57465
156 466	c	NO	P	NO	NH	52466	57466
156 467	r	FS	A	SR	CK	52467	57467
156 468	d	NO	A	NO	NH	52468	57468
156 469	d	NO	A	NO	HT	52469	57469
156 470	d	NO	A	NO	NH	52470	57470
156 471	d	NO	A	NO	NH	52471	57471
156 472	d	NO	A	NO	NH	52472	57472
156 473	d	NO	A	NO	NH	52473	57473
156 474	rt	FS	A	SR	CK	52474	57474
156 475	d	NO	A	NO	HT	52475	57475
156 476	rt	FS	A	SR	CK	52476	57476
156 477	t	FS	A	SR	CK	52477	57477
156 478	rt	FS	A	SR	CK	52478	57478
156 479	d	NO	A	NO	HT	52479	57479
156 480	d	NO	A	NO	HT	52480	57480
156 481	d	NO	A	NO	HT	52481	57481
156 482	d	NO	A	NO	NH	52482	57482
156 483	d	NO	A	NO	NH	52483	57483
156 484	d	NO	A	NO	HT	52484	57484
156 485	ru	FS	A	SR	CK	52485	57485
156 486	d	NO	A	NO	NH	52486	57486
156 487	d	NO	A	NO	NH	52487	57487
156 488	d	NO	A	NO	NH	52488	57488
156 489	d	NO	A	NO	NH	52489	57489

156 490	d	**NO**	A	*NO*	HT	52490 57490
156 491	d	**NO**	A	*NO*	NH	52491 57491
156 492	rt	**FS**	A	*SR*	CK	52492 57492
156 493	rt	**FS**	A	*SR*	CK	52493 57493
156 494	ru	**SR**	A	*SR*	CK	52494 57494
156 495	u	**SR**	A	*SR*	CK	52495 57495
156 496	ru	**FS**	A	*SR*	CK	52496 57496
156 497	d	**NO**	A	*NO*	NH	52497 57497
156 498	d	**NO**	A	*NO*	NH	52498 57498
156 499	rt	**FS**	A	*SR*	CK	52499 57499
156 500	u	**SR**	A	*SR*	CK	52500 57500
156 501		**SR**	A	*SR*	CK	52501 57501
156 502		**SR**	A	*SR*	CK	52502 57502
156 503		**SR**	A	*SR*	CK	52503 57503
156 504		**SR**	A	*SR*	CK	52504 57504
156 505		**SR**	A	*SR*	CK	52505 57505
156 506		**SR**	A	*SR*	CK	52506 57506
156 507		**SR**	A	*SR*	CK	52507 57507
156 508		**SR**	A	*SR*	CK	52508 57508
156 509		**SR**	A	*SR*	CK	52509 57509
156 510		**SR**	A	*SR*	CK	52510 57510
156 511		**SR**	A	*SR*	CK	52511 57511
156 512		**SR**	A	*SR*	CK	52512 57512
156 513		**SR**	A	*SR*	CK	52513 57513
156 514		**SR**	A	*SR*	CK	52514 57514

Names:

156 409	Cromer Pier Seaside Special
156 416	Saint Edmund
156 420	LA' AL RATTY Ravenglass & Eskdale Railway
156 441	William Huskisson MP
156 444	Councillor Bill Cameron
156 459	Benny Rothman – The Manchester Rambler
156 460	Driver John Axon G.C.
156 464	Lancashire DalesRail
156 466	Gracie Fields

CLASS 158/0   BREL

DMSL(B)–DMSL(A) or DMCL–DMSL or DMSL–MSL–DMSL.

Construction: Welded aluminium.
Engines: 158 701–158 813/158 880–158 890/158 950–158 959: One Cummins
NTA855R of 260 kW (350 h.p.) at 1900 r.p.m.
158 815–158 862: One Perkins 2006-TWH of 260 kW (350 h.p.) at 1900 r.p.m.
158 863–158 872: One Cummins NTA855R of 300 kW (400 h.p.) at 2100 r.p.m.
Bogies: One BREL P4 (powered) and one BREL T4 (non-powered) per car.
Couplers: BSI.
Dimensions: 22.57 x 2.70 m.
Gangways: Throughout.   **Wheel Arrangement:** 2-B + B-2.
Doors: Twin-leaf swing plug.   **Maximum Speed:** 90 m.p.h.

Seating Layout: 2+2 facing/unidirectional in all Standard and First Class except 2+1 in South West Trains First Class.

Multiple Working: Within class and with Classes 142, 143, 144, 150, 153, 155, 156, 159, 170 and 172.

DMSL(B). Lot No. 31051 BREL Derby 1989–1992. –/68 1TD 1W. († –/72 1TD 1W, c, w –/66 1TD 1W, t –/64 1TD 1W). 38.5 t.

MSL. Lot No. 31050 BREL Derby 1991. –/66(3) 1T. 38.5 t.

DMSL(A). Lot No. 31052 BREL Derby 1989–1992. –/70 1T († –/74, c, w –/68 1T, * –/64(2) 1T plus cycle stowage area, t –/66 1T). 38.5 t.

The above details refer to the "as built" condition. The following DMSL(B) have now been converted to DMCL as follows:

52701–52736/52738–52741 (ScotRail). 15/53 1TD 1W (* refurbished sets 14/46(6) 1TD 1W plus cycle stowage area).

52786/52789 (Former South West Trains units). 13/44 1TD 1W.

Northern promotional vinyls:

158 784 PTEG: 40 years.
158 787, 158 792–796 Sheffield–Leeds fast service.
158 790 Rugby League (Northern Rail Cup).
158 860 Keighley & Brontë Country.
158 901–910 Leeds–Bradford–Manchester route (the "Calder Valley").

Notes:

*   Refurbished ScotRail units fitted with Grammer seating, additional luggage racks and cycle stowage areas.
    ScotRail units 158 726–741 are fitted with Richmond seating.
†   Refurbished East Midlands Trains units with Primarius seating.
c   Chapman seating.
s   Arriva Trains Wales and Northern units with some seats removed for additional luggage space.
t   Refurbished former South West Trains units with Class 159-style interiors, including First Class seating.
w   Refurbished First Great Western units. Units 158 745–751 & 158 762 (most formed into 3-car sets) have been fitted with Richmond seating.

All ScotRail 158s are "fitted" for RETB. When a unit arrives at Inverness the cab display unit is clipped on and plugged in. Similarly Arriva Trains Wales units have RETB plugged in at Shrewsbury for working the Cambrian Lines.

158 701	*	**FS**	P	*SR*	IS	52701	57701
158 702	*	**FS**	P	*SR*	IS	52702	57702
158 703	*	**FS**	P	*SR*	IS	52703	57703
158 704	*	**FS**	P	*SR*	IS	52704	57704
158 705	*	**FS**	P	*SR*	IS	52705	57705
158 706	*	**FS**	P	*SR*	IS	52706	57706
158 707	*	**FS**	P	*SR*	IS	52707	57707
158 708	*	**FS**	P	*SR*	IS	52708	57708
158 709	*	**FS**	P	*SR*	IS	52709	57709
158 710	*	**FS**	P	*SR*	IS	52710	57710
158 711	*	**FS**	P	*SR*	IS	52711	57711
158 712	*	**FS**	P	*SR*	IS	52712	57712

158 713	*	**FS**	P	*SR*	IS	52713	57713	
158 714	*	**FS**	P	*SR*	IS	52714	57714	
158 715	*	**FS**	P	*SR*	IS	52715	57715	
158 716	*	**FS**	P	*SR*	IS	52716	57716	
158 717	*	**FS**	P	*SR*	IS	52717	57717	
158 718	*	**FS**	P	*SR*	IS	52718	57718	
158 719	*	**FS**	P	*SR*	IS	52719	57719	
158 720	*	**FS**	P	*SR*	IS	52720	57720	
158 721	*	**FS**	P	*SR*	IS	52721	57721	
158 722	*	**FS**	P	*SR*	IS	52722	57722	
158 723	*	**FS**	P	*SR*	IS	52723	57723	
158 724	*	**FS**	P	*SR*	IS	52724	57724	
158 725	*	**FS**	P	*SR*	IS	52725	57725	
158 726		**FS**	P	*SR*	HA	52726	57726	
158 727		**FS**	P	*SR*	HA	52727	57727	
158 728		**FS**	P	*SR*	HA	52728	57728	
158 729		**FS**	P	*SR*	HA	52729	57729	
158 730		**FS**	P	*SR*	HA	52730	57730	
158 731		**FS**	P	*SR*	HA	52731	57731	
158 732		**FS**	P	*SR*	HA	52732	57732	
158 733		**FS**	P	*SR*	HA	52733	57733	
158 734		**FS**	P	*SR*	HA	52734	57734	
158 735		**FS**	P	*SR*	HA	52735	57735	
158 736		**FS**	P	*SR*	HA	52736	57736	
158 738		**FS**	P	*SR*	HA	52738	57738	
158 739		**FS**	P	*SR*	HA	52739	57739	
158 740		**FS**	P	*SR*	HA	52740	57740	
158 741		**FS**	P	*SR*	HA	52741	57741	
158 749	w	**FI**	P	*GW*	PM	52749	57749	
158 752		**N0**	P	*NO*	NL	52752	58716	57752
158 753		**N0**	P	*NO*	NL	52753	58710	57753
158 754		**N0**	P	*NO*	NL	52754	58708	57754
158 755		**N0**	P	*NO*	NL	52755	58702	57755
158 756		**N0**	P	*NO*	NL	52756	58712	57756
158 757		**N0**	P	*NO*	NL	52757	58706	57757
158 758		**N0**	P	*NO*	NL	52758	58714	57758
158 759		**N0**	P	*NO*	NL	52759	58713	57759
158 763	w	**FI**	P	*GW*	PM	52763	57763	
158 766	w	**FI**	P	*GW*	PM	52766	57766	
158 767	w	**FI**	P	*GW*	PM	52767	57767	
158 769	w	**FI**	P	*GW*	PM	52769	57769	
158 770	†	**ST**	P	*EM*	NM	52770	57770	
158 773	†	**ST**	P	*EM*	NM	52773	57773	
158 774	†	**ST**	P	*EM*	NM	52774	57774	
158 777	†	**ST**	P	*EM*	NM	52777	57777	
158 780	†	**ST**	A	*EM*	NM	52780	57780	
158 782		**FB**	A	*SR*	HA	52782	57782	
158 783	†	**ST**	A	*EM*	NM	52783	57783	
158 784	t	**N0**	A	*NO*	NL	52784	57784	
158 785	†	**ST**	A	*EM*	NM	52785	57785	
158 786	u	**FB**	A	*SR*	HA	52786	57786	

58 787		NO	A	NO	NL	52787	57787
58 788	†	ST	A	EM	NM	52788	57788
58 789	u	FB	A	SR	HA	52789	57789
58 790	t	NO	A	NO	NL	52790	57790
58 791	t	NO	A	NO	NL	52791	57791
58 792		NO	A	NO	NL	52792	57792
58 793		NO	A	NO	NL	52793	57793
58 794		NO	A	NO	NL	52794	57794
58 795		NO	A	NO	NL	52795	57795
58 796		NO	A	NO	NL	52796	57796
58 797	t	NO	A	NO	NL	52797	57797
58 798	w	FI	P	GW	PM	52798	58715	57798
58 799	†	ST	P	EM	NM	52799	57799
58 806	†	ST	P	EM	NM	52806	57806
58 810	†	ST	P	EM	NM	52810	57810
58 812	†	ST	P	EM	NM	52812	57812
58 813	†	ST	P	EM	NM	52813	57813
58 815	c	NO	A	NO	NL	52815	57815
58 816	c	NO	A	NO	NL	52816	57816
58 817	c	NO	A	NO	NL	52817	57817
58 818	ce	AV	A	AW	MN	52818	57818
58 819	ce	WB	A	AW	MN	52819	57819
58 820	ce	AV	A	AW	MN	52820	57820
58 821	ce	AV	A	AW	MN	52821	57821
58 822	ce	AV	A	AW	MN	52822	57822
58 823	ce	AV	A	AW	MN	52823	57823
58 824	ce	AV	A	AW	MN	52824	57824
58 825	ce	WB	A	AW	MN	52825	57825
58 826	ce	WB	A	AW	MN	52826	57826
58 827	ce	WB	A	AW	MN	52827	57827
58 828	ce	AV	A	AW	MN	52828	57828
58 829	ce	AV	A	AW	MN	52829	57829
58 830	ce	WB	A	AW	MN	52830	57830
58 831	ce	WB	A	AW	MN	52831	57831
58 832	ce	WB	A	AW	MN	52832	57832
58 833	ce	WB	A	AW	MN	52833	57833
58 834	ce	WB	A	AW	MN	52834	57834
58 835	ce	WB	A	AW	MN	52835	57835
58 836	ce	WB	A	AW	MN	52836	57836
58 837	ce	AV	A	AW	MN	52837	57837
58 838	ce	WB	A	AW	MN	52838	57838
58 839	ce	WB	A	AW	MN	52839	57839
58 840	ce	AV	A	AW	MN	52840	57840
58 841	ce	WB	A	AW	MN	52841	57841
58 842	c	NO	A	NO	NL	52842	57842
58 843	c	NO	A	NO	NL	52843	57843
58 844	t	NO	A	NO	NL	52844	57844
58 845	t	NO	A	NO	NL	52845	57845
58 846	†	ST	A	EM	NM	52846	57846
58 847	†	ST	A	EM	NM	52847	57847
58 848	t	NO	A	NO	NL	52848	57848

158 849	t	**NO**	A	*NO*	NL	52849	57849
158 850	t	**NO**	A	*NO*	NL	52850	57850
158 851	t	**NO**	A	*NO*	NL	52851	57851
158 852	†	**ST**	A	*EM*	NM	52852	57852
158 853	t	**NO**	A	*NO*	NL	52853	57853
158 854	†	**ST**	A	*EM*	NM	52854	57854
158 855		**NO**	A	*NO*	NL	52855	57855
158 856	†	**ST**	A	*EM*	NM	52856	57856
158 857	†	**ST**	A	*EM*	NM	52857	57857
158 858	†	**ST**	A	*EM*	NM	52858	57858
158 859		**NO**	A	*NO*	NL	52859	57859
158 860		**NO**	A	*NO*	NL	52860	57860
158 861		**NO**	A	*NO*	NL	52861	57861
158 862	†	**ST**	A	*EM*	NM	52862	57862
158 863	†	**ST**	A	*EM*	NM	52863	57863
158 864	†	**ST**	A	*EM*	NM	52864	57864
158 865	†	**ST**	A	*EM*	NM	52865	57865
158 866	†	**ST**	A	*EM*	NM	52866	57866
158 867	c	**SR**	A	*SR*	HA	52867	57867
158 868	c	**SR**	A	*SR*	HA	52868	57868
158 869	c	**WT**	A	*SR*	HA	52869	57869
158 870	c	**SR**	A	*SR*	HA	52870	57870
158 871	c	**SR**	A	*SR*	HA	52871	57871
158 872	c	**NO**	A	*NO*	NL	52872	57872

Names:

158 702	BBC Scotland 75 years
158 707	Far North Line 125th ANNIVERSARY
158 715	Haymarket
158 720	Inverness & Nairn Railway – 150 years
158 784	Barbara Castle
158 791	County of Nottinghamshire
158 796	Fred Trueman Cricketing Legend
158 860	Ian Dewhirst

Class 158/8. Refurbished South West Trains units. Converted from former TransPennine Express units at Wabtec, Doncaster in 2007. 2+1 seating in First Class. Details as Class 158/0 except:

DMCL. Lot No. 31051 BREL Derby 1989–1992. 13/44 1TD 1W. 38.5 t.
DMSL. Lot No. 31052 BREL Derby 1989–1992. –/70 1T. 38.5 t.

158 880	(158 737)	**ST**	P	*SW*	SA	52737	57737
158 881	(158 742)	**ST**	P	*SW*	SA	52742	57742
158 882	(158 743)	**ST**	P	*SW*	SA	52743	57743
158 883	(158 744)	**ST**	P	*SW*	SA	52744	57744
158 884	(158 772)	**ST**	P	*SW*	SA	52772	57772
158 885	(158 775)	**ST**	P	*SW*	SA	52775	57775
158 886	(158 779)	**ST**	P	*SW*	SA	52779	57779
158 887	(158 781)	**ST**	P	*SW*	SA	52781	57781
158 888	(158 802)	**ST**	P	*SW*	SA	52802	57802
158 889	(158 808)	**ST**	P	*SW*	SA	52808	57808
158 890	(158 814)	**ST**	P	*SW*	SA	52814	57814

CLASS 158/9   BREL

DMSL–DMS. Units leased by West Yorkshire PTE but managed by Eversholt Rail. Details as Class 158/0 except for seating and toilets.

DMSL. Lot No. 31051 BREL Derby 1990–1992. –/70 1TD 1W. 38.5 t.
DMS. Lot No. 31052 BREL Derby 1990–1992. –/72 and parcels area. 38.5 t.

158 901	**NO**	E	*NO*	NL	52901	57901	
158 902	**NO**	E	*NO*	NL	52902	57902	
158 903	**NO**	E	*NO*	NL	52903	57903	
158 904	**NO**	E	*NO*	NL	52904	57904	
158 905	**NO**	E	*NO*	NL	52905	57905	
158 906	**NO**	E	*NO*	NL	52906	57906	
158 907	**NO**	E	*NO*	NL	52907	57907	
158 908	**NO**	E	*NO*	NL	52908	57908	
158 909	**NO**	E	*NO*	NL	52909	57909	
158 910	**NO**	E	*NO*	NL	52910	57910	William Wilberforce

CLASS 158/0   BREL

DMSL–DMSL–DMSL. Refurbished units reformed in 2008 for First Great Western. For vehicle details see above. Formations can be flexible depending on when unit exams become due.

158 950	w	**FI**	P	*GW*	PM	57751	52761	57761
158 951	w	**FI**	P	*GW*	PM	57751	52764	57764
158 952	w	**FI**	P	*GW*	PM	57745	52762	57762
158 953	w	**FI**	P	*GW*	PM	57745	52750	57750
158 954	w	**FI**	P	*GW*	PM	57747	52760	57760
158 955	w	**FI**	P	*GW*	PM	57747	52765	57765
158 956	w	**FI**	P	*GW*	PM	57748	52768	57768
158 957	w	**FI**	P	*GW*	PM	57748	52771	57771
158 958	w	**FI**	P	*GW*	PM	57746	52776	57776
158 959	w	**FI**	P	*GW*	PM	52746	52778	57778

CLASS 159/0   BREL

DMCL–MSL–DMSL. Built as Class 158. Converted before entering passenger service to Class 159 by Rosyth Dockyard.

Construction: Welded aluminium.
Engines: One Cummins NTA855R of 300 kW (400 h.p.) at 2100 r.p.m.
Bogies: One BREL P4 (powered) and one BREL T4 (non-powered) per car.
Couplers: BSI.
Dimensions: 22.16 x 2.70 m.
Gangways: Throughout.   **Wheel Arrangement:** 2-B + B-2 + B-2.
Doors: Twin-leaf swing plug.   **Maximum Speed:** 90 m.p.h.
Seating Layout: 1: 2+1 facing, 2: 2+2 facing/unidirectional.
Multiple Working: Within class and with Classes 142, 143, 144, 150, 153, 155, 156, 158 and 170.

DMCL. Lot No. 31051 BREL Derby 1992–1993. 23/28 1TD 1W. 38.5 t.
MSL. Lot No. 31050 BREL Derby 1992–1993. –/70(6) 1T. 38.5 t.
DMSL. Lot No. 31052 BREL Derby 1992–1993. –/72 1T. 38.5 t.

159 001	**ST**	P	*SW*	SA	52873	58718	57873	CITY OF EXETER
159 002	**ST**	P	*SW*	SA	52874	58719	57874	CITY OF SALISBURY
159 003	**ST**	P	*SW*	SA	52875	58720	57875	TEMPLECOMBE
159 004	**ST**	P	*SW*	SA	52876	58721	57876	BASINGSTOKE AND DEANE
159 005	**ST**	P	*SW*	SA	52877	58722	57877	
159 006	**ST**	P	*SW*	SA	52878	58723	57878	
159 007	**ST**	P	*SW*	SA	52879	58724	57879	
159 008	**ST**	P	*SW*	SA	52880	58725	57880	
159 009	**ST**	P	*SW*	SA	52881	58726	57881	
159 010	**ST**	P	*SW*	SA	52882	58727	57882	
159 011	**ST**	P	*SW*	SA	52883	58728	57883	
159 012	**ST**	P	*SW*	SA	52884	58729	57884	
159 013	**ST**	P	*SW*	SA	52885	58730	57885	
159 014	**ST**	P	*SW*	SA	52886	58731	57886	
159 015	**ST**	P	*SW*	SA	52887	58732	57887	
159 016	**ST**	P	*SW*	SA	52888	58733	57888	
159 017	**ST**	P	*SW*	SA	52889	58734	57889	
159 018	**ST**	P	*SW*	SA	52890	58735	57890	
159 019	**ST**	P	*SW*	SA	52891	58736	57891	
159 020	**ST**	P	*SW*	SA	52892	58737	57892	
159 021	**ST**	P	*SW*	SA	52893	58738	57893	
159 022	**ST**	P	*SW*	SA	52894	58739	57894	

CLASS 159/1   BREL

DMCL–MSL–DMSL. Units converted from Class 158s at Wabtec, Doncaster in 2006–07 for South West Trains.

Details as Class 158/0 except:
Seating Layout: 1: 2+1 facing, 2: 2+2 facing/unidirectional.

DMCL. Lot No. 31051 BREL Derby 1989–1992. 24/28 1TD 1W. 38.5 t.
MSL. Lot No. 31050 BREL Derby 1989–1992. –/70 1T. 38.5 t.
DMSL. Lot No. 31052 BREL Derby 1989–1992. –/72 1T.38.5 t.

159 101	(158 800)	**ST**	P	*SW*	SA	52800	58717	57800
159 102	(158 803)	**ST**	P	*SW*	SA	52803	58703	57803
159 103	(158 804)	**ST**	P	*SW*	SA	52804	58704	57804
159 104	(158 805)	**ST**	P	*SW*	SA	52805	58705	57805
159 105	(158 807)	**ST**	P	*SW*	SA	52807	58707	57807
159 106	(158 809)	**ST**	P	*SW*	SA	52809	58709	57809
159 107	(158 811)	**ST**	P	*SW*	SA	52811	58711	57811
159 108	(158 801)	**ST**	P	*SW*	SA	52801	58701	57801

CLASS 165/0   NETWORK TURBO   BREL

DMSL–DMS and DMSL–MS–DMS. Chiltern Railways units. Refurbished 2003–2005 with First Class seats removed and air conditioning fitted.

Construction: Welded aluminium.
Engines: One Perkins 2006-TWH of 260 kW (350 h.p.) at 1900 r.p.m.
Bogies: BREL P3-17 (powered), BREL T3-17 (non-powered).
Couplers: BSI.
Dimensions: 23.50/23.25 x 2.81 m.
Gangways: Within unit only.   **Wheel Arrangement:** 2-B (+ B-2) + B-2.
Doors: Twin-leaf swing plug.   **Maximum Speed:** 75 m.p.h.
Seating Layout: 2+2/3+2 facing/unidirectional.
Multiple Working: Within class and with Classes 166 and 168.

Fitted with tripcocks for working over London Underground tracks between Harrow-on-the-Hill and Amersham.

58801–58822/58873–58878. DMSL. Lot No. 31087 BREL York 1990. –/82(7) 1T 2W. 40.1 t.
58823–58833. DMSL. Lot No. 31089 BREL York 1991–1992. –/82(7) 1T 2W. 40.1 t.
MS. Lot No. 31090 BREL York 1991–1992. –/106. 37.0 t.
DMS. Lot No. 31088 BREL York 1991–1992. –/94. 39.4 t.

165 001	**CR**	A	*CR*	AL	58801		58834
165 002	**CR**	A	*CR*	AL	58802		58835
165 003	**CR**	A	*CR*	AL	58803		58836
165 004	**CR**	A	*CR*	AL	58804		58837
165 005	**CR**	A	*CR*	AL	58805		58838
165 006	**CR**	A	*CR*	AL	58806		58839
165 007	**CR**	A	*CR*	AL	58807		58840
165 008	**CR**	A	*CR*	AL	58808		58841
165 009	**CR**	A	*CR*	AL	58809		58842
165 010	**CR**	A	*CR*	AL	58810		58843
165 011	**CR**	A	*CR*	AL	58811		58844
165 012	**CR**	A	*CR*	AL	58812		58845
165 013	**CR**	A	*CR*	AL	58813		58846
165 014	**CR**	A	*CR*	AL	58814		58847
165 015	**CR**	A	*CR*	AL	58815		58848
165 016	**CR**	A	*CR*	AL	58816		58849
165 017	**CR**	A	*CR*	AL	58817		58850
165 018	**CR**	A	*CR*	AL	58818		58851
165 019	**CR**	A	*CR*	AL	58819		58852
165 020	**CR**	A	*CR*	AL	58820		58853
165 021	**CR**	A	*CR*	AL	58821		58854
165 022	**CR**	A	*CR*	AL	58822		58855
165 023	**CR**	A	*CR*	AL	58873		58867
165 024	**CR**	A	*CR*	AL	58874		58868
165 025	**CR**	A	*CR*	AL	58875		58869
165 026	**CR**	A	*CR*	AL	58876		58870
165 027	**CR**	A	*CR*	AL	58877		58871
165 028	**CR**	A	*CR*	AL	58878		58872
165 029	**CR**	A	*CR*	AL	58823	55404	58856
165 030	**CR**	A	*CR*	AL	58824	55405	58857
165 031	**CR**	A	*CR*	AL	58825	55406	58858
165 032	**CR**	A	*CR*	AL	58826	55407	58859
165 033	**CR**	A	*CR*	AL	58827	55408	58860

165 034	**CR**	A	*CR*	AL	58828	55409	58861
165 035	**CR**	A	*CR*	AL	58829	55410	58862
165 036	**CR**	A	*CR*	AL	58830	55411	58863
165 037	**CR**	A	*CR*	AL	58831	55412	58864
165 038	**CR**	A	*CR*	AL	58832	55413	58865
165 039	**CR**	A	*CR*	AL	58833	55414	58866

CLASS 165/1   NETWORK TURBO   BREL

First Great Western units. DMCL–MS–DMS or DMCL–DMS.

Construction: Welded aluminium.
Engines: One Perkins 2006-TWH of 260 kW (350 h.p.) at 1900 r.p.m.
Bogies: BREL P3-17 (powered), BREL T3-17 (non-powered).
Couplers: BSI.
Dimensions: 23.50/23.25 x 2.81 m.
Gangways: Within unit only.   **Wheel Arrangement:** 2-B (+ B-2) + B-2.
Doors: Twin-leaf swing plug.   **Maximum Speed:** 90 m.p.h.
Seating Layout: 1: 2+2 facing, 2: 3+2 facing/unidirectional.
Multiple Working: Within class and with Classes 166 and 168.

58953–58969. DMCL. Lot No. 31098 BREL York 1992. 16/66 1T. 38.0 t.
58879–58898. DMCL. Lot No. 31096 BREL York 1992. 16/72 1T. 38.0 t.
MS. Lot No. 31099 BREL 1992. –/106. 37.0 t.
DMS. Lot No. 31097 BREL 1992. –/98. 37.0 t.

165 101	**FD**	A	*GW*	RG	58953	55415	58916
165 102	**FD**	A	*GW*	RG	58954	55416	58917
165 103	**FD**	A	*GW*	RG	58955	55417	58918
165 104	**FD**	A	*GW*	RG	58956	55418	58919
165 105	**FD**	A	*GW*	RG	58957	55419	58920
165 106	**FD**	A	*GW*	RG	58958	55420	58921
165 107	**FD**	A	*GW*	RG	58959	55421	58922
165 108	**FD**	A	*GW*	RG	58960	55422	58923
165 109	**FD**	A	*GW*	RG	58961	55423	58924
165 110	**FD**	A	*GW*	RG	58962	55424	58925
165 111	**FD**	A	*GW*	RG	58963	55425	58926
165 112	**FD**	A	*GW*	RG	58964	55426	58927
165 113	**FD**	A	*GW*	RG	58965	55427	58928
165 114	**FD**	A	*GW*	RG	58966	55428	58929
165 116	**FD**	A	*GW*	RG	58968	55430	58931
165 117	**FD**	A	*GW*	RG	58969	55431	58932
165 118	**FD**	A	*GW*	RG	58879		58933
165 119	**FD**	A	*GW*	RG	58880		58934
165 120	**FD**	A	*GW*	RG	58881		58935
165 121	**FD**	A	*GW*	RG	58882		58936
165 122	**FD**	A	*GW*	RG	58883		58937
165 123	**FD**	A	*GW*	RG	58884		58938
165 124	**FD**	A	*GW*	RG	58885		58939
165 125	**FD**	A	*GW*	RG	58886		58940
165 126	**FD**	A	*GW*	RG	58887		58941
165 127	**FD**	A	*GW*	RG	58888		58942
165 128	**FD**	A	*GW*	RG	58889		58943

165 129	FD	A	GW	RG	58890	58944
165 130	FD	A	GW	RG	58891	58945
165 131	FD	A	GW	RG	58892	58946
165 132	FD	A	GW	RG	58893	58947
165 133	FD	A	GW	RG	58894	58948
165 134	FD	A	GW	RG	58895	58949
165 135	FD	A	GW	RG	58896	58950
165 136	FD	A	GW	RG	58897	58951
165 137	FD	A	GW	RG	58898	58952

CLASS 166   NETWORK EXPRESS TURBO   ABB

DMCL(A)–MS–DMCL(B). First Great Western units, built for Paddington–Oxford/
Newbury services. Air conditioned and with additional luggage space compared
to the Class 165s.

Construction: Welded aluminium.
Engines: One Perkins 2006-TWH of 260 kW (350 h.p.) at 1900 r.p.m.
Bogies: BREL P3-17 (powered), BREL T3-17 (non-powered).
Couplers: BSI.
Dimensions: 23.50 x 2.81 m.
Gangways: Within unit only.   **Wheel Arrangement:** 2-B + B-2 + B-2.
Doors: Twin-leaf swing plug.   **Maximum Speed:** 90 m.p.h.
Seating Layout: 1: 2+2 facing, 2: 2+2/3+2 facing/unidirectional.
Multiple Working: Within class and with Classes 165 and 168.

DMCL (A). Lot No. 31116 ABB York 1992–1993. 16/68 1T. 39.6 t.
MS. Lot No. 31117 ABB York 1992–1993. –/91. 38.0 t.
DMCL (B). Lot No. 31116 ABB York 1992–1993. 16/68 1T. 39.6 t.

166 201	FD	A	GW	RG	58101	58601	58122
166 202	FD	A	GW	RG	58102	58602	58123
166 203	FD	A	GW	RG	58103	58603	58124
166 204	FD	A	GW	RG	58104	58604	58125
166 205	FD	A	GW	RG	58105	58605	58126
166 206	FD	A	GW	RG	58106	58606	58127
166 207	FD	A	GW	RG	58107	58607	58128
166 208	FD	A	GW	RG	58108	58608	58129
166 209	FD	A	GW	RG	58109	58609	58130
166 210	FD	A	GW	RG	58110	58610	58131
166 211	FD	A	GW	RG	58111	58611	58132
166 212	FD	A	GW	RG	58112	58612	58133
166 213	FD	A	GW	RG	58113	58613	58134
166 214	FD	A	GW	RG	58114	58614	58135
166 215	FD	A	GW	RG	58115	58615	58136
166 216	FD	A	GW	RG	58116	58616	58137
166 217	FD	A	GW	RG	58117	58617	58138
166 218	FD	A	GW	RG	58118	58618	58139
166 219	FD	A	GW	RG	58119	58619	58140
166 220	FD	A	GW	RG	58120	58620	58141
166 221	FD	A	GW	RG	58121	58621	58142

CLASS 168   CLUBMAN   ADTRANZ/BOMBARDIER

Air conditioned.

Construction: Welded aluminium bodies with bolt-on steel ends.
Engines: One MTU 6R183TD13H of 315 kW (422 h.p.) at 1900 r.p.m.
Transmission: Hydraulic. Voith T211rzze to ZF final drive.
Bogies: One Adtranz P3–23 and one BREL T3–23 per car.
Couplers: BSI at outer ends, bar within unit.
Dimensions: Class 168/0: 24.1/23.61 x 2.69 m. Others: 23.62/23.61 x 2.69 m.
Gangways: Within unit only.   **Wheel Arrangement:** 2-B (+ B-2 + B-2) + B-2.
Doors: Twin-leaf swing plug.   **Maximum Speed:** 100 m.p.h.
Seating Layout: 2+2 facing/unidirectional.
Multiple Working: Within class and with Classes 165 and 166.

Fitted with tripcocks for working over London Underground tracks between
Harrow-on-the-Hill and Amersham.

Class 168/0. Original Design. DMSL(A)–MS–MSL–DMSL(B) or DMSL(A)–MSL–
MS–DMSL(B).

58151–58155. DMSL(A). Adtranz Derby 1997–1998. –/57 1TD 1W. 44.0 t.
58651–58655. MSL. Adtranz Derby 1998. –/73 1T. 41.0 t.
58451–58455. MS. Adtranz Derby 1998. –/77. 41.0 t.
58251–58255. DMSL(B). Adtranz Derby 1998. –/68 1T. 43.6 t.

Note: 58451–58455 were numbered 58656–58660 for a time when used in
168 106–168 110.

168 001	**CR**	P	*CR*	AL	58151	58451	58651	58251
168 002	**CR**	P	*CR*	AL	58152	58652	58452	58252
168 003	**CR**	P	*CR*	AL	58153	58653	58453	58253
168 004	**CR**	P	*CR*	AL	58154	58654	58454	58254
168 005	**CR**	P	*CR*	AL	58155	58655	58455	58255

Class 168/1. These units are effectively Class 170s. DMSL(A)–MSL–MS–DMSL(B)
or DMSL(A)–MS–DMSL(B).

58156–58163. DMSL(A). Adtranz Derby 2000. –/57 1TD 2W. 45.2 t.
58456–58460. MS. Bombardier Derby 2002. –/76. 41.8 t.
58756–58757. MSL. Bombardier Derby 2002. –/73 1T. 42.9 t.
58461–58463. MS. Adtranz Derby 2000. –/76. 42.4 t.
58256–58263. DMSL(B). Adtranz Derby 2000. –/69 1T. 45.2 t.

Notes: 58461–58463 have been renumbered from 58661–58663.

168 106	**CR**	P	*CR*	AL	58156	58756	58456	58256
168 107	**CR**	P	*CR*	AL	58157	58757	58457	58257
168 108	**CR**	P	*CR*	AL	58158		58458	58258
168 109	**CR**	P	*CR*	AL	58159		58459	58259
168 110	**CR**	P	*CR*	AL	58160		58460	58260
168 111	**CR**	E	*CR*	AL	58161		58461	58261
168 112	**CR**	E	*CR*	AL	58162		58462	58262
168 113	**CR**	E	*CR*	AL	58163		58463	58263

Class 168/2. These units are effectively Class 170s. DMSL(A)–(MS)–MS–DMSL(B).

58164–58169. DMSL(A). Bombardier Derby 2003–2004. –/57 1TD 2W. 45.4 t.
58365–58367. MS. Bombardier Derby 2006. –/76. 43.3 t.
58464/58468/58469. MS. Bombardier Derby 2003–2004. –/76. 44.0 t.
58465–58467. MS. Bombardier Derby 2006. –/76. 43.3 t.
58264–58269. DMSL(B). Bombardier Derby 2003–2004. –/69 1T. 45.5 t.

168 214	**CR**	P	*CR*	AL	58164		58464	58264
168 215	**CR**	P	*CR*	AL	58165	58465	58365	58265
168 216	**CR**	P	*CR*	AL	58166	58366	58466	58266
168 217	**CR**	P	*CR*	AL	58167	58367	58467	58267
168 218	**CR**	P	*CR*	AL	58168		58468	58268
168 219	**CR**	P	*CR*	AL	58169		58469	58269

CLASS 170   TURBOSTAR   ADTRANZ/BOMBARDIER

Various formations. Air conditioned.

Construction: Welded aluminium bodies with bolt-on steel ends.
Engines: One MTU 6R183TD13H of 315 kW (422 h.p.) at 1900 r.p.m.
Transmission: Hydraulic. Voith T211rzze to ZF final drive.
Bogies: One Adtranz P3–23 and one BREL T3–23 per car.
Couplers: BSI at outer ends, bar within later build units.
Dimensions: 23.62/23.61 x 2.69 m.
Gangways: Within unit only.   **Wheel Arrangement:** 2-B (+ B-2) + B-2.
Doors: Twin-leaf sliding plug.   **Maximum Speed:** 100 m.p.h.
Seating Layout: 1: 2+1 facing/unidirectional. 2: 2+2 unidirectional/facing.
Multiple Working: Within class and with Classes 150, 153, 155, 156, 158, 159 and 172.

Class 170/1. CrossCountry (former Midland Mainline) units. Lazareni seating. DMSL–MS–DMCL/DMSL–DMCL.

DMSL. Adtranz Derby 1998–1999. –/59 1TD 2W. 45.0 t.
MS. Adtranz Derby 2001. –/80. 43.0 t.
DMCL. Adtranz Derby 1998–1999. 9/52 1T. 44.8 t

170 101	**XC**	P	*XC*	TS	50101	55101	79101
170 102	**XC**	P	*XC*	TS	50102	55102	79102
170 103	**XC**	P	*XC*	TS	50103	55103	79103
170 104	**XC**	P	*XC*	TS	50104	55104	79104
170 105	**XC**	P	*XC*	TS	50105	55105	79105
170 106	**XC**	P	*XC*	TS	50106	55106	79106
170 107	**XC**	P	*XC*	TS	50107	55107	79107
170 108	**XC**	P	*XC*	TS	50108	55108	79108
170 109	**XC**	P	*XC*	TS	50109	55109	79109
170 110	**XC**	P	*XC*	TS	50110	55110	79110
170 111	**XC**	P	*XC*	TS	50111		79111
170 112	**XC**	P	*XC*	TS	50112		79112
170 113	**XC**	P	*XC*	TS	50113		79113
170 114	**XC**	P	*XC*	TS	50114		79114
170 115	**XC**	P	*XC*	TS	50115		79115

| 170 116 | **XC** | P | *XC* | TS | 50116 | | 79116 |
| 170 117 | **XC** | P | *XC* | TS | 50117 | | 79117 |

Class 170/2. National Express East Anglia 3-car units. Chapman seating. DMCL–MSL–DMSL.

DMCL. Adtranz Derby 1999. 7/39 1TD 2W. 45.0 t.
MSL. Adtranz Derby 1999. –/68 1T. Guard's office. 45.3 t.
DMSL. Adtranz Derby 1999. –/66 1T. 43.4 t.

170 201	r	**1**	P	*EA*	NC	50201	56201	79201
170 202	r	**1**	P	*EA*	NC	50202	56202	79202
170 203	r	**1**	P	*EA*	NC	50203	56203	79203
170 204	r	**1**	P	*EA*	NC	50204	56204	79204
170 205	r	**1**	P	*EA*	NC	50205	56205	79205
170 206	r	**1**	P	*EA*	NC	50206	56206	79206
170 207	r	**1**	P	*EA*	NC	50207	56207	79207
170 208	r	**1**	P	*EA*	NC	50208	56208	79208

Class 170/2. National Express East Anglia 2-car units. Chapman seating. DMSL–DMCL.

DMSL. Bombardier Derby 2002. –/57 1TD 2W. 45.7 t.
DMCL. Bombardier Derby 2002. 9/53 1T. 45.7 t.

170 270	r	**1**	P	*EA*	NC	50270	79270
170 271	r	**AN**	P	*EA*	NC	50271	79271
170 272	r	**AN**	P	*EA*	NC	50272	79272
170 273	r	**AN**	P	*EA*	NC	50273	79273

Class 170/3. TransPennine Express units. Chapman seating. DMCL–DMSL. 170 309 renumbered from 170 399.

50301–50308/50399. DMCL. Adtranz Derby 2000–2001. 8/43 1TD 2W. 45.8 t.
79301–79308/79399. DMSL. Adtranz Derby 2000–2001. –/65 1T. 45.8 t.

170 301	**FT**	P	*TP*	XW	50301	79301
170 302	**FT**	P	*TP*	XW	50302	79302
170 303	**FT**	P	*TP*	XW	50303	79303
170 304	**FT**	P	*TP*	XW	50304	79304
170 305	**FT**	P	*TP*	XW	50305	79305
170 306	**FT**	P	*TP*	XW	50306	79306
170 307	**FT**	P	*TP*	XW	50307	79307
170 308	**FT**	P	*TP*	XW	50308	79308
170 309	**FT**	P	*TP*	XW	50399	79399

Class 170/3. Units built for Hull Trains, now in use with ScotRail. Chapman seating. DMCL–MSLRB–DMSL.

DMCL. Bombardier Derby 2004. 7/41 1TD 2W. 46.5 t.
MSLRB. Bombardier Derby 2004. –/53 1T. Buffet and guard's office 44.7 t.
DMSL. Bombardier Derby 2004. –/67 1T. 46.3 t.

170 393	**FS**	P	*SR*	HA	50393	56393	79393
170 394	**FS**	P	*SR*	HA	50394	56394	79394
170 395	**FS**	P	*SR*	HA	50395	56395	79395
170 396	**FS**	P	*SR*	HA	50396	56396	79396

Class 170/3. CrossCountry units. Lazareni seating. DMSL–MS–DMCL.

DMSL. Bombardier Derby 2002. –/59 1TD 2W. 45.4 t.
MS. Bombardier Derby 2002. –/80. 43.0 t.
DMCL. Bombardier Derby 2002. 9/52 1T. 45.8 t.

170 397	**XC**	P	*XC*	TS	50397	56397	79397
170 398	**XC**	P	*XC*	TS	50398	56398	79398

Class 170/4. ScotRail "express" units. Chapman seating. DMCL–MS–DMCL.

DMCL(A). Adtranz Derby 1999–2001. 9/43 1TD 2W. 45.2 t.
MS. Adtranz Derby 1999–2001. –/76. 42.5 t.
DMCL(B). Adtranz Derby 1999–2001. 9/49 1T. 45.2 t.

170 401	**FS**	P	*SR*	HA	50401	56401	79401
170 402	**FS**	P	*SR*	HA	50402	56402	79402
170 403	**FS**	P	*SR*	HA	50403	56403	79403
170 404	**FS**	P	*SR*	HA	50404	56404	79404
170 405	**FS**	P	*SR*	HA	50405	56405	79405
170 406	**FS**	P	*SR*	HA	50406	56406	79406
170 407	**FS**	P	*SR*	HA	50407	56407	79407
170 408	**FS**	P	*SR*	HA	50408	56408	79408
170 409	**FS**	P	*SR*	HA	50409	56409	79409
170 410	**FS**	P	*SR*	HA	50410	56410	79410
170 411	**FS**	P	*SR*	HA	50411	56411	79411
170 412	**FS**	P	*SR*	HA	50412	56412	79412
170 413	**FS**	P	*SR*	HA	50413	56413	79413
170 414	**FS**	P	*SR*	HA	50414	56414	79414
170 415	**FS**	P	*SR*	HA	50415	56415	79415
170 416	**FS**	H	*SR*	HA	50416	56416	79416
170 417	**FS**	H	*SR*	HA	50417	56417	79417
170 418	**FS**	H	*SR*	HA	50418	56418	79418
170 419	**FS**	H	*SR*	HA	50419	56419	79419
170 420	**FS**	H	*SR*	HA	50420	56420	79420
170 421	**FS**	H	*SR*	HA	50421	56421	79421
170 422	**FS**	H	*SR*	HA	50422	56422	79422
170 423	**FS**	H	*SR*	HA	50423	56423	79423
170 424	**FS**	H	*SR*	HA	50424	56424	79424

Class 170/4. ScotRail "express" units. Chapman seating. DMCL–MS–DMCL.

DMCL. Bombardier Derby 2003–2005. 9/43 1TD 2W. 46.8 t.
MS. Bombardier Derby 2003–2005. –/76. 43.7 t.
DMCL. Bombardier Derby 2003–2005. 9/49 1T. 46.5 t.

Note: * 170 431 & 170 432 have new uprated engines fitted: MTU 6H1800R83 of 360 kW (483 h.p.) at 1800 r.p.m.

170 425	**FS**	P	*SR*	HA	50425	56425	79425
170 426	**FS**	P	*SR*	HA	50426	56426	79426
170 427	**FS**	P	*SR*	HA	50427	56427	79427
170 428	**FS**	P	*SR*	HA	50428	56428	79428
170 429	**FS**	P	*SR*	HA	50429	56429	79429
170 430	**FS**	P	*SR*	HA	50430	56430	79430

170 431	*	**FS**	P	*SR*	HA	50431	56431	79431	
170 432	*	**FS**	P	*SR*	HA	50432	56432	79432	
170 433		**FS**	P	*SR*	HA	50433	56433	79433	Investor in People
170 434		**SR**	P	*SR*	HA	50434	56434	79434	

Class 170/4. ScotRail units. Originally built as Standard Class only units. 170 450–455 retro-fitted with First Class in 2008. Chapman seating. DMSL–MS–DMSL or † DMCL–MS–DMCL.

DMSL. Bombardier Derby 2004–2005. –/55 1TD 2W († 9/47 1TD 2W). 46.3 t.
MS. Bombardier Derby 2004–2005. –/76. 43.4 t.
DMSL. Bombardier Derby 2004–2005. –/67 1T († 9/49 1T 1W). 46.4 t.

170 450	†	**FS**	P	*SR*	HA	50450	56450	79450
170 451	†	**FS**	P	*SR*	HA	50451	56451	79451
170 452	†	**FS**	P	*SR*	HA	50452	56452	79452
170 453	†	**FS**	P	*SR*	HA	50453	56453	79453
170 454	†	**FS**	P	*SR*	HA	50454	56454	79454
170 455	†	**FS**	P	*SR*	HA	50455	56455	79455
170 456		**FS**	P	*SR*	HA	50456	56456	79456
170 457		**FS**	P	*SR*	HA	50457	56457	79457
170 458		**FS**	P	*SR*	HA	50458	56458	79458
170 459		**FS**	P	*SR*	HA	50459	56459	79459
170 460		**FS**	P	*SR*	HA	50460	56460	79460
170 461		**FS**	P	*SR*	HA	50461	56461	79461

Class 170/4. ScotRail units. Standard Class only units. Chapman seating. DMSL–MS–DMSL.

50470–50471. DMSL(A). Adtranz Derby 2001. –/55 1TD 2W. 45.1 t.
50472–50478. DMSL(A). Bombardier Derby 2004–2005. –/57 1TD 2W. 46.3 t.
56470–56471. MS. Adtranz Derby 2001. –/76. 42.4 t.
56472–56478. MS. Bombardier Derby 2004–2005. –/76. 43.4 t.
79470–79471. DMSL(B). Adtranz Derby 2001. –/67 1T. 45.1 t.
79472–79478. DMSL(B). Bombardier Derby 2004–2005. –/67 1T. 46.4 t.

170 470	**SC**	P	*SR*	HA	50470	56470	79470
170 471	**SC**	P	*SR*	HA	50471	56471	79471
170 472	**SP**	P	*SR*	HA	50472	56472	79472
170 473	**SP**	P	*SR*	HA	50473	56473	79473
170 474	**SP**	P	*SR*	HA	50474	56474	79474
170 475	**SP**	P	*SR*	HA	50475	56475	79475
170 476	**SP**	P	*SR*	HA	50476	56476	79476
170 477	**SP**	P	*SR*	HA	50477	56477	79477
170 478	**SP**	P	*SR*	HA	50478	56478	79478

Class 170/5. London Midland and CrossCountry 2-car units. Lazareni seating. DMSL–DMSL or * DMSL–DMCL (CrossCountry).

DMSL(A). Adtranz Derby 1999–2000. –/55 1TD 2W (* –/59 1TD 2W). 45.8 t.
DMSL(B). Adtranz Derby 1999–2000. –/67 1T (* DMCL 9/52 1T). 45.9 t.

170 501	**LM**	P	*LM*	TS	50501	79501
170 502	**LM**	P	*LM*	TS	50502	79502
170 503	**LM**	P	*LM*	TS	50503	79503

170 504	**LM**	P	*LM*	TS	50504	79504
170 505	**LM**	P	*LM*	TS	50505	79505
170 506	**LM**	P	*LM*	TS	50506	79506
170 507	**LM**	P	*LM*	TS	50507	79507
170 508	**LM**	P	*LM*	TS	50508	79508
170 509	**LM**	P	*LM*	TS	50509	79509
170 510	**LM**	P	*LM*	TS	50510	79510
170 511	**LM**	P	*LM*	TS	50511	79511
170 512	**LM**	P	*LM*	TS	50512	79512
170 513	**LM**	P	*LM*	TS	50513	79513
170 514	**LM**	P	*LM*	TS	50514	79514
170 515	**LM**	P	*LM*	TS	50515	79515
170 516	**LM**	P	*LM*	TS	50516	79516
170 517	**LM**	P	*LM*	TS	50517	79517
170 518	* **XC**	P	*XC*	TS	50518	79518
170 519	* **XC**	P	*XC*	TS	50519	79519
170 520	* **XC**	P	*XC*	TS	50520	79520
170 521	* **XC**	P	*XC*	TS	50521	79521
170 522	* **XC**	P	*XC*	TS	50522	79522
170 523	* **XC**	P	*XC*	TS	50523	79523

Class 170/6. London Midland and CrossCountry 3-car units. Lazareni seating. DMSL–MS–DMSL or * DMSL–MS–DMCL (Cross-Country).

DMSL(A). Adtranz Derby 2000. –/55 1TD 2W (* –/59 1TD 2W). 45.8 t.
MS. Adtranz Derby 2000. –/74 (* –/80). 42.4 t.
DMSL(B). Adtranz Derby 2000. –/67 1T (* DMCL 9/52 1T). 45.9 t.

170 630	**LM**	P	*LM*	TS	50630	56630	79630
170 631	**LM**	P	*LM*	TS	50631	56631	79631
170 632	**LM**	P	*LM*	TS	50632	56632	79632
170 633	**LM**	P	*LM*	TS	50633	56633	79633
170 634	**LM**	P	*LM*	TS	50634	56634	79634
170 635	**LM**	P	*LM*	TS	50635	56635	79635
170 636	* **XC**	P	*XC*	TS	50636	56636	79636
170 637	* **XC**	P	*XC*	TS	50637	56637	79637
170 638	* **XC**	P	*XC*	TS	50638	56638	79638
170 639	* **XC**	P	*XC*	TS	50639	56639	79639

CLASS 171   TURBOSTAR   BOMBARDIER

DMCL–DMSL or DMCL–MS–MS–DMCL. Southern units. Air conditioned. Chapman seating.

Construction: Welded aluminium bodies with bolt-on steel ends.
Engines: One MTU 6R183TD13H of 315 kW (422 h.p.) at 1900 r.p.m.
Transmission: Hydraulic. Voith T211rzze to ZF final drive.
Bogies: One Adtranz P3–23 and one BREL T3–23 per car.
Couplers: Dellner 12 at outer ends, bar within unit (Class 171/8s).
Dimensions: 23.62/23.61 x 2.69 m.
Gangways: Within unit only.   **Wheel Arrangement:** 2-B (+ B-2 + B-2) + B-2.
Doors: Twin-leaf swing plug.   **Maximum Speed:** 100 m.p.h.

Seating Layout: 1: 2+1 facing/unidirectional. 2: 2+2 facing/unidirectional.
Multiple Working: Within class and with EMU Classes 375 and 377 in an emergency.

Class 171/7. 2-car units. DMCL–DMSL.

50721–50726. DMCL. Bombardier Derby 2003. 9/43 1TD 2W. 47.6 t.
50727–50729. DMCL. Bombardier Derby 2005. 9/43 1TD 2W. 46.3 t.
50392. DMCL. Bombardier Derby 2003. 9/43 1TD 2W. 46.6 t.
79721–79726. DMSL. Bombardier Derby 2003. –/64 1T. 47.8 t.
79727–79729. DMSL. Bombardier Derby 2005. –/64 1T. 46.2 t.
79392. DMSL. Bombardier Derby 2003. –/64 1T. 46.5 t.

Notes: 171 721–726 were built as Class 170s (170 721–726), but renumbered as Class 171s on fitting with Dellner couplers.
171 730 was formerly South West Trains unit 170 392, before transferring to Southern in 2007.

171 721	**SN**	P	*SN*	SU	50721	79721
171 722	**SN**	P	*SN*	SU	50722	79722
171 723	**SN**	P	*SN*	SU	50723	79723
171 724	**SN**	P	*SN*	SU	50724	79724
171 725	**SN**	P	*SN*	SU	50725	79725
171 726	**SN**	P	*SN*	SU	50726	79726
171 727	**SN**	P	*SN*	SU	50727	79727
171 728	**SN**	P	*SN*	SU	50728	79728
171 729	**SN**	P	*SN*	SU	50729	79729
171 730	**SN**	P	*SN*	SU	50392	79392

Class 171/8. 4-car units. DMCL(A)–MS–MS–DMCL(B).

DMCL(A). Bombardier Derby 2004. 9/43 1TD 2W. 46.5 t.
MS. Bombardier Derby 2004. –/74. 43.7 t.
DMCL(B). Bombardier Derby 2004. 9/50 1T. 46.5 t.

171 801	**SN**	P	*SN*	SU	50801	54801	56801	79801
171 802	**SN**	P	*SN*	SU	50802	54802	56802	79802
171 803	**SN**	P	*SN*	SU	50803	54803	56803	79803
171 804	**SN**	P	*SN*	SU	50804	54804	56804	79804
171 805	**SN**	P	*SN*	SU	50805	54805	56805	79805
171 806	**SN**	P	*SN*	SU	50806	54806	56806	79806

CLASS 172   TURBOSTAR   BOMBARDIER

New generation Turbostars on order for London Overground, Chiltern Railways and London Midland. Air conditioned. Full details awaited (those shown for Chiltern and London Midland units are provisional).

Construction: Welded aluminium bodies with bolt-on steel ends.
Engines: One MTU 6H1800R83 of 360 kW (483 h.p.) at 1800 r.p.m.
Transmission: Mechanical. Supplied by ZG, Germany.
Bogies: B5006 type "lightweight" bogies.
Couplers: BSI at outer ends, bar within unit.
Dimensions: 23.0 x 2.69 m.

Gangways: London Overground & Chiltern units: Within unit only. London Midland units: Throughout.
Wheel Arrangement:
Doors: Twin-leaf sliding plug.
Maximum Speed: 75 m.p.h. (London Midland units 100 m.p.h.)
Seating Layout: 2+2 facing/unidirectional.
Multiple Working: Within class and with Classes 150, 153, 155, 156, 158, 159 and 170.

Class 172/0. London Overground units. Used on Gospel Oak–Barking line. DMS–DMS.

59311–59318. DMS(W). Bombardier Derby 2009–2010. –/60 2W. 41.6 t.
59411–59418. DMS. Bombardier Derby 2009–2010. –/64. 41.5 t.

172 001	LO	A		WN	59311 59411
172 002	LO	A		WN	59312 59412
172 003	LO	A		WN	59313 59413
172 004	LO	A	LO	WN	59314 59414
172 005	LO	A	LO	WN	59315 59415
172 006	LO	A	LO	WN	59316 59416
172 007	LO	A		WN	59317 59417
172 008	LO	A			59318 59418

Class 172/1. Chiltern Railways units. DMSL–DMS. On order. Due for delivery from spring 2011.

59111–59114. DMSL. Bombardier Derby 2009–2010. –/65 1TD 2W. 41.4 t.
59211–59214. DMS. Bombardier Derby 2009–2010. –/80. 40.8 t.

172 101	A	59111	59211
172 102	A	59112	59212
172 103	A	59113	59213
172 104	A	59114	59214

Class 172/2. London Midland 2-car units. DMSL–DMS. On order for use on West Midlands suburban services. Due for delivery from spring 2011.

50211–50222. DMSL. Bombardier Derby 2009–2010. –/53(4) 1TD 2W. 41.9 t.
79211–79222. DMS. Bombardier Derby 2009–2010. –/68(3). 41.3 t.

172 211	P	50211	79211
172 212	P	50212	79212
172 213	P	50213	79213
172 214	P	50214	79214
172 215	P	50215	79215
172 216	P	50216	79216
172 217	P	50217	79217
172 218	P	50218	79218
172 219	P	50219	79219
172 220	P	50220	79220
172 221	P	50221	79221
172 222	P	50222	79222

Class 172/3. London Midland 3-car units. DMSL–MS–DMS. On order for use on West Midlands suburban services. Due for delivery from spring 2011.

50331–50345. DMSL. Bombardier Derby 2009–2010. –/53(4) 1TD 2W. 41.9 t.
56331–56345. MS. Bombardier Derby 2009–2010. –/72. 38.1 t.
79331–79345. DMS. Bombardier Derby 2009–2010. –/68(3). 41.3. t.

172 331	P	50331	56331	79331
172 332	P	50332	56332	79332
172 333	P	50333	56333	79333
172 334	P	50334	56334	79334
172 335	P	50335	56335	79335
172 336	P	50336	56336	79336
172 337	P	50337	56337	79337
172 338	P	50338	56338	79338
172 339	P	50339	56339	79339
172 340	P	50340	56340	79340
172 341	P	50341	56341	79341
172 342	P	50342	56342	79342
172 343	P	50343	56343	79343
172 344	P	50344	56344	79344
172 345	P	50345	56345	79345

CLASS 175   CORADIA 1000   ALSTOM

Air conditioned.

Construction: Steel.
Engines: One Cummins N14 of 335 kW (450 h.p.).
Transmission: Hydraulic. Voith T211rzze to ZF Voith final drive.
Bogies: ACR (Alstom FBO) – LTB-MBS1, TB-MB1, MBS1-LTB.
Couplers: Scharfenberg outer ends and bar within unit (Class 175/1).
Dimensions: 23.7 x 2.73 m.
Gangways: Within unit only.   **Wheel Arrangement:** 2-B (+ B-2) + B-2.
Doors: Single-leaf swing plug.   **Maximum Speed:** 100 m.p.h.
Seating Layout: 2+2 facing/unidirectional.
Multiple Working: Within class and with Class 180.

Class 175/0. DMSL–DMSL. 2-car units.

DMSL(A). Alstom Birmingham 1999–2000. –/54 1TD 2W. 48.8 t.
DMSL(B). Alstom Birmingham 1999–2000. –/64 1T. 50.7 t.

175 001	**AV**	A	*AW*	CH	50701	79701
175 002	**AV**	A	*AW*	CH	50702	79702
175 003	**AV**	A	*AW*	CH	50703	79703
175 004	**AV**	A	*AW*	CH	50704	79704
175 005	**AV**	A	*AW*	CH	50705	79705
175 006	**AV**	A	*AW*	CH	50706	79706
175 007	**AV**	A	*AW*	CH	50707	79707
175 008	**AV**	A	*AW*	CH	50708	79708
175 009	**AV**	A	*AW*	CH	50709	79709
175 010	**AV**	A	*AW*	CH	50710	79710
175 011	**AV**	A	*AW*	CH	50711	79711

Class 175/1. DMSL–MSL–DMSL. 3-car units.

DMSL(A). Alstom Birmingham 1999–2001. –/54 1TD 2W. 50.7 t.
MSL. Alstom Birmingham 1999–2001. –/68 1T. 47.5 t.
DMSL(B). Alstom Birmingham 1999–2001. –/64 1T. 49.5 t.

175 101	**AV**	A	*AW*	CH	50751	56751	79751
175 102	**AV**	A	*AW*	CH	50752	56752	79752
175 103	**AV**	A	*AW*	CH	50753	56753	79753
175 104	**AV**	A	*AW*	CH	50754	56754	79754
175 105	**AV**	A	*AW*	CH	50755	56755	79755
175 106	**AV**	A	*AW*	CH	50756	56756	79756
175 107	**AV**	A	*AW*	CH	50757	56757	79757
175 108	**AV**	A	*AW*	CH	50758	56758	79758
175 109	**AV**	A	*AW*	CH	50759	56759	79759
175 110	**AV**	A	*AW*	CH	50760	56760	79760
175 111	**AV**	A	*AW*	CH	50761	56761	79761
175 112	**AV**	A	*AW*	CH	50762	56762	79762
175 113	**AV**	A	*AW*	CH	50763	56763	79763
175 114	**AV**	A	*AW*	CH	50764	56764	79764
175 115	**AV**	A	*AW*	CH	50765	56765	79765
175 116	**AV**	A	*AW*	CH	50766	56766	79766

CLASS 180   CORADIA 1000   ALSTOM

Air conditioned.

Construction: Steel.
Engines: One Cummins QSK19 of 560 kW (750 h.p.) at 2100 r.p.m.
Transmission: Hydraulic. Voith T312br to Voith final drive.
Bogies: ACR (Alstom FBO): LTB1-MBS2, TB1-MB2, TB1-MB2, TB2-MB2, MBS2-LTB1.
Couplers: Scharfenberg outer ends, bar within unit.
Dimensions: 23.71/23.03 x 2.73 m.
Gangways: Within unit only.
Wheel Arrangement: 2-B + B-2 + B-2 + B-2 + B-2.
Doors: Single-leaf swing plug.   **Maximum Speed:** 125 m.p.h.
Seating Layout: 1: 2+1 facing/unidirectional. 2: 2+2 facing/unidirectional.
Multiple Working: Within class and with Class 175.

DMSL(A). Alstom Birmingham 2000–2001. –/46 2W 1TD. 51.7 t.
MFL. Alstom Birmingham 2000–2001. 42/– 1T 1W + catering point. 49.6 t.
MSL. Alstom Birmingham 2000–2001. –/68 1T. 49.5 t.
MSLRB. Alstom Birmingham 2000–2001. –/56 1T. 50.3 t.
DMSL(B). Alstom Birmingham 2000–2001. –/56 1T. 51.4 t.

180 101	**GC**	A	*GC*	HT	50901	54901	55901	56901	59901
180 102	**FG**	A	*HT*	OO	50902	54902	55902	56902	59902
180 103	**FB**	A	*NO*	NH	50903	54903	55903	56903	59903
180 104	**FG**	A		OO	50904	54904	55904	56904	59904
180 105	**GC**	A	*GC*	HT	50905	54905	55905	56905	59905
180 106	**FB**	A	*NO*	NH	50906	54906	55906	56906	59906
180 107	**GC**	A	*GC*	HT	50907	54907	55907	56907	59907
180 108	**FB**	A	*NO*	NH	50908	54908	55908	56908	59908

180 109	**FG**	A	*HT*	OO	50909	54909	55909	56909	59909
180 110	**FD**	A	*HT*	OO	50910	54910	55910	56910	59910
180 111	**FD**	A	*HT*	OO	50911	54911	55911	56911	59911
180 112	**GC**	A	*GC*	HT	50912	54912	55912	56912	59912
180 113	**FD**	A	*HT*	OO	50913	54913	55913	56913	59913
180 114	**GC**	A	*GC*	HT	50914	54914	55914	56914	59914

Name (carried on DMSL(A)): 180 112   JAMES HERRIOT

CLASS 185   DESIRO UK   SIEMENS

Air conditioned. Grammer seating.

Construction: Aluminium.
Engines: One Cummins QSK19 of 560 kW (750 h.p.) at 2100 r.p.m.
Transmission: Voith.
Bogies: Siemens.
Couplers: Dellner 12.
Dimensions: 23.76/23.75 x 2.66 m.
Gangways: Within unit only.   **Wheel Arrangement:** 2-B + 2-B + B-2.
Doors: Double-leaf sliding plug.   **Maximum Speed:** 100 m.p.h.
Seating Layout: 1: 2+1 facing/unidirectional, 2: 2+2 facing/unidirectional.
Multiple Working: Within class only.

DMCL. Siemens Uerdingen 2005–2006. 15/18(8) 2W 1TD + catering point. 55.4 t.
MSL. Siemens Uerdingen 2005–2006. –/72 1T. 52.7 t.
DMS. Siemens Uerdingen 2005–2006. –/64(4). 54.9 t.

185 101	**FT**	E	*TP*	AK	51101	53101	54101
185 102	**FT**	E	*TP*	AK	51102	53102	54102
185 103	**FT**	E	*TP*	AK	51103	53103	54103
185 104	**FT**	E	*TP*	AK	51104	53104	54104
185 105	**FT**	E	*TP*	AK	51105	53105	54105
185 106	**FT**	E	*TP*	AK	51106	53106	54106
185 107	**FT**	E	*TP*	AK	51107	53107	54107
185 108	**FT**	E	*TP*	AK	51108	53108	54108
185 109	**FT**	E	*TP*	AK	51109	53109	54109
185 110	**FT**	E	*TP*	AK	51110	53110	54110
185 111	**FT**	E	*TP*	AK	51111	53111	54111
185 112	**FT**	E	*TP*	AK	51112	53112	54112
185 113	**FT**	E	*TP*	AK	51113	53113	54113
185 114	**FT**	E	*TP*	AK	51114	53114	54114
185 115	**FT**	E	*TP*	AK	51115	53115	54115
185 116	**FT**	E	*TP*	AK	51116	53116	54116
185 117	**FT**	E	*TP*	AK	51117	53117	54117
185 118	**FT**	E	*TP*	AK	51118	53118	54118
185 119	**FT**	E	*TP*	AK	51119	53119	54119
185 120	**FT**	E	*TP*	AK	51120	53120	54120
185 121	**FT**	E	*TP*	AK	51121	53121	54121
185 122	**FT**	E	*TP*	AK	51122	53122	54122
185 123	**FT**	E	*TP*	AK	51123	53123	54123
185 124	**FT**	E	*TP*	AK	51124	53124	54124

185 125	FT	E	TP	AK	51125	53125	54125
185 126	FT	E	TP	AK	51126	53126	54126
185 127	FT	E	TP	AK	51127	53127	54127
185 128	FT	E	TP	AK	51128	53128	54128
185 129	FT	E	TP	AK	51129	53129	54129
185 130	FT	E	TP	AK	51130	53130	54130
185 131	FT	E	TP	AK	51131	53131	54131
185 132	FT	E	TP	AK	51132	53132	54132
185 133	FT	E	TP	AK	51133	53133	54133
185 134	FT	E	TP	AK	51134	53134	54134
185 135	FT	E	TP	AK	51135	53135	54135
185 136	FT	E	TP	AK	51136	53136	54136
185 137	FT	E	TP	AK	51137	53137	54137
185 138	FT	E	TP	AK	51138	53138	54138
185 139	FT	E	TP	AK	51139	53139	54139
185 140	FT	E	TP	AK	51140	53140	54140
185 141	FT	E	TP	AK	51141	53141	54141
185 142	FT	E	TP	AK	51142	53142	54142
185 143	FT	E	TP	AK	51143	53143	54143
185 144	FT	E	TP	AK	51144	53144	54144
185 145	FT	E	TP	AK	51145	53145	54145
185 146	FT	E	TP	AK	51146	53146	54146
185 147	FT	E	TP	AK	51147	53147	54147
185 148	FT	E	TP	AK	51148	53148	54148
185 149	FT	E	TP	AK	51149	53149	54149
185 150	FT	E	TP	AK	51150	53150	54150
185 151	FT	E	TP	AK	51151	53151	54151

2. DIESEL ELECTRIC UNITS

CLASS 201/202 PRESERVED "HASTINGS" UNIT BR

DMBS–TSL–TSL–TSRB–TSL–DMBS.

Preserved unit made up from two Class 201 short-frame cars and three Class 202 long-frame cars. The "Hastings" units were made with narrow body-profiles for use on the section between Tonbridge and Battle which had tunnels of restricted loading gauge. These tunnels were converted to single track operation in the 1980s thus allowing standard loading gauge stock to be used. The set also contains a Class 411 EMU trailer (not Hastings line gauge) and a Class 422 EMU buffet car.

Construction: Steel.
Engine: One English Electric 4SRKT Mk. 2 of 450 kW (600 h.p.) at 850 r.p.m.
Main Generator: English Electric EE824.
Traction Motors: Two English Electric EE507 mounted on the inner bogie.
Bogies: SR Mk. 4. (Former EMU TSL vehicles have Commonwealth bogies).
Couplers: Drophead buckeye.

Dimensions: 18.40 x 2.50 m (60000), 20.35 x 2.50 m (60116/60118/60529), 18.36 x 2.50 m (60501), 20.35 x 2.82 (69337), 20.30 x 2.82 (70262).
Gangways: Within unit only.
Doors: Manually operated slam.
Brakes: Electro-pneumatic and automatic air.
Maximum Speed: 75 m.p.h.
Seating Layout: 2+2 facing.
Multiple Working: Other ex-BR Southern Region DEMU vehicles.

60000. DMBS. Lot No. 30329 Eastleigh 1957. –/22. 55.0 t.
60116. DMBS. Lot No. 30395 Eastleigh 1957. –/31. 56.0 t.
60118. DMBS. Lot No. 30395 Eastleigh 1957. –/30. 56.0 t.
60501. TSL. Lot No. 30331 Eastleigh 1957. –/52 2T. 29.5 t.
60529. TSL. Lot No. 30397 Eastleigh 1957. –/60 2T. 30.5 t.
69337. TSRB (ex-Class 422 EMU). Lot No. 30805 York 1970. –/40. 35.0 t.
70262. TSL (ex-Class 411/5 EMU). Lot No. 30455 Eastleigh 1958. –/64 2T. 31.5 t.

201 001	**G**	HD *HD*	SE	60116	60529	70262	69337	60501	60118
Spare	**G**	HD *HD*	SE	60000					

Names:

60000 Hastings
60116 Mountfield
60118 Tunbridge Wells

CLASS 220   VOYAGER   BOMBARDIER

DMS–MS–MS–DMF.

Construction: Steel.
Engine: Cummins QSK19 of 560 kW (750 h.p.) at 1800 r.p.m.
Transmission: Two Alstom Onix 800 three-phase traction motors of 275 kW.
Braking: Rheostatic and electro-pneumatic.
Bogies: Bombardier B5005.
Couplers: Dellner 12 at outer ends, bar within unit.
Dimensions: 23.85/23.00(602xx) x 2.73 m.
Gangways: Within unit only.
Wheel Arrangement: 1A-A1 + 1A-A1 + 1A-A1 + 1A-A1.
Doors: Single-leaf swing plug.
Maximum Speed: 125 m.p.h.
Seating Layout: 1: 2+1 facing/unidirectional, 2: 2+2 mainly unidirectional.
Multiple Working: Within class and with Classes 221 and 222 (in an emergency). Also can be controlled from Class 57/3 locomotives.

DMS. Bombardier Brugge/Wakefield 2000–2001. –/42 1TD 1W. 51.1 t.
MS (A). Bombardier Brugge/Wakefield 2000–2001. –/66. 45.9 t.
MS (B). Bombardier Brugge/Wakefield 2000–2001. –/66 1TD. 46.7 t.
DMF. Bombardier Brugge/Wakefield 2000–2001. 26/– 1TD 1W. 50.9 t.

220 001	**XC**	HX *XC*	CZ	60301	60701	60201	60401
220 002	**XC**	HX *XC*	CZ	60302	60702	60202	60402
220 003	**XC**	HX *XC*	CZ	60303	60703	60203	60403
220 004	**XC**	HX *XC*	CZ	60304	60704	60204	60404

220 005	**XC**	HX	*XC*	CZ	60305	60705	60205	60405
220 006	**XC**	HX	*XC*	CZ	60306	60706	60206	60406
220 007	**XC**	HX	*XC*	CZ	60307	60707	60207	60407
220 008	**XC**	HX	*XC*	CZ	60308	60708	60208	60408
220 009	**XC**	HX	*XC*	CZ	60309	60709	60209	60409
220 010	**XC**	HX	*XC*	CZ	60310	60710	60210	60410
220 011	**XC**	HX	*XC*	CZ	60311	60711	60211	60411
220 012	**XC**	HX	*XC*	CZ	60312	60712	60212	60412
220 013	**XC**	HX	*XC*	CZ	60313	60713	60213	60413
220 014	**XC**	HX	*XC*	CZ	60314	60714	60214	60414
220 015	**XC**	HX	*XC*	CZ	60315	60715	60215	60415
220 016	**XC**	HX	*XC*	CZ	60316	60716	60216	60416
220 017	**XC**	HX	*XC*	CZ	60317	60717	60217	60417
220 018	**XC**	HX	*XC*	CZ	60318	60718	60218	60418
220 019	**XC**	HX	*XC*	CZ	60319	60719	60219	60419
220 020	**XC**	HX	*XC*	CZ	60320	60720	60220	60420
220 021	**XC**	HX	*XC*	CZ	60321	60721	60221	60421
220 022	**XC**	HX	*XC*	CZ	60322	60722	60222	60422
220 023	**XC**	HX	*XC*	CZ	60323	60723	60223	60423
220 024	**XC**	HX	*XC*	CZ	60324	60724	60224	60424
220 025	**XC**	HX	*XC*	CZ	60325	60725	60225	60425
220 026	**XC**	HX	*XC*	CZ	60326	60726	60226	60426
220 027	**XC**	HX	*XC*	CZ	60327	60727	60227	60427
220 028	**XC**	HX	*XC*	CZ	60328	60728	60228	60428
220 029	**XC**	HX	*XC*	CZ	60329	60729	60229	60429
220 030	**XC**	HX	*XC*	CZ	60330	60730	60230	60430
220 031	**XC**	HX	*XC*	CZ	60331	60731	60231	60431
220 032	**XC**	HX	*XC*	CZ	60332	60732	60232	60432
220 033	**XC**	HX	*XC*	CZ	60333	60733	60233	60433
220 034	**XC**	HX	*XC*	CZ	60334	60734	60234	60434

CLASS 221   SUPER VOYAGER   BOMBARDIER

* DMS–MS–(MS)–MSRMB–DMF (Virgin Trains units) or DMS–MS–MS–MS–DMF (CrossCountry units). Built as tilting units but tilt now isolated on CrossCountry sets.

Construction: Steel.
Engine: Cummins QSK19 of 560 kW (750 h.p.) at 1800 r.p.m.
Transmission: Two Alstom Onix 800 three-phase traction motors of 275 kW.
Braking: Rheostatic and electro-pneumatic.
Bogies: Bombardier HVP.
Couplers: Dellner 12 at outer ends, bar within unit.
Dimensions: 23.67 x 2.73 m.
Gangways: Within unit only.
Wheel Arrangement: 1A-A1 + 1A-A1 + 1A-A1 (+ 1A-A1) + 1A-A1.
Doors: Single-leaf swing plug.
Maximum Speed: 125 m.p.h.
Seating Layout: 1: 2+1 facing/unidirectional, 2: 2+2 mainly unidirectional.
Multiple Working: Within class and with Classes 220 and 222 (in an emergency). Also can be controlled from Class 57/3 locomotives.

DMS. Bombardier Brugge/Wakefield 2001–2002. –/42 1TD 1W. 58.5 t (* 58.9 t.).
60751–794 MS (* MSRMB). Bombardier Brugge/Wakefield 2001–2002. –/66 (* –/52). 54.1 t (* 55.9 t.).
60951–994. MS. Bombardier Brugge/Wakefield 2001–2002. –/66 1TD (* –/68 1TD). 54.8 t (* 54.3 t.).
60851–890. MS. Bombardier Brugge/Wakefield 2001–2002. –/62 1TD (* –/68 1TD). 54.4 t (* 55.0 t.).
DMF. Bombardier Brugge/Wakefield 2001–2002. 26/– 1TD 1W. 58.9 t (* 59.1 t.).

Advertising livery: 221 115 Dark grey Bombardier branding on end vehicles.

Note: * Virgin Trains units. MSRMB moved adjacent to the DMF. The seating in this vehicle (2+2 facing) can be used by First or Standard Class passengers depending on demand.

221 101	*	**VT**	HX	*VW*	CZ	60351	60951	60851	60751	60451
221 102	*	**VT**	HX	*VW*	CZ	60352	60952	60852	60752	60452
221 103	*	**VT**	HX	*VW*	CZ	60353	60953	60853	60753	60453
221 104	*	**VT**	HX	*VW*	CZ	60354	60954	60854	60754	60454
221 105	*	**VT**	HX	*VW*	CZ	60355	60955	60855	60755	60455
221 106	*	**VT**	HX	*VW*	CZ	60356	60956	60856	60756	60456
221 107	*	**VT**	HX	*VW*	CZ	60357	60957	60857	60757	60457
221 108	*	**VT**	HX	*VW*	CZ	60358	60958	60858	60758	60458
221 109	*	**VT**	HX	*VW*	CZ	60359	60959	60859	60759	60459
221 110	*	**VT**	HX	*VW*	CZ	60360	60960	60860	60760	60460
221 111	*	**VT**	HX	*VW*	CZ	60361	60961	60861	60761	60461
221 112	*	**VT**	HX	*VW*	CZ	60362	60962	60862	60762	60462
221 113	*	**VT**	HX	*VW*	CZ	60363	60963	60863	60763	60463
221 114	*	**VT**	HX	*VW*	CZ	60364	60764	60964	60864	60464
221 115	*	**AL**	HX	*VW*	CZ	60365	60765	60965	60865	60465
221 116	*	**VT**	HX	*VW*	CZ	60366	60766	60966	60866	60466
221 117	*	**VT**	HX	*VW*	CZ	60367	60767	60967	60867	60467
221 118	*	**VT**	HX	*VW*	CZ	60368	60768	60968	60868	60468
221 119		**XC**	HX	*XC*	CZ	60369	60769	60969	60869	60469
221 120		**XC**	HX	*XC*	CZ	60370	60770	60970	60870	60470
221 121		**XC**	HX	*XC*	CZ	60371	60771	60971	60871	60471
221 122		**XC**	HX	*XC*	CZ	60372	60772	60972	60872	60472
221 123		**XC**	HX	*XC*	CZ	60373	60773	60973	60873	60473
221 124		**XC**	HX	*XC*	CZ	60374	60774	60974	60874	60474
221 125		**XC**	HX	*XC*	CZ	60375	60775	60975	60875	60475
221 126		**XC**	HX	*XC*	CZ	60376	60776	60976	60876	60476
221 127		**XC**	HX	*XC*	CZ	60377	60777	60977	60877	60477
221 128		**XC**	HX	*XC*	CZ	60378	60778	60978	60878	60478
221 129		**XC**	HX	*XC*	CZ	60379	60779	60979	60879	60479
221 130		**XC**	HX	*XC*	CZ	60380	60780	60980	60880	60480
221 131		**XC**	HX	*XC*	CZ	60381	60781	60981	60881	60481
221 132		**XC**	HX	*XC*	CZ	60382	60782	60982	60882	60482
221 133		**XC**	HX	*XC*	CZ	60383	60783	60983	60883	60483
221 134		**XC**	HX	*XC*	CZ	60384	60784	60984	60884	60484
221 135		**XC**	HX	*XC*	CZ	60385	60785	60985	60885	60485
221 136		**XC**	HX	*XC*	CZ	60386	60786	60986	60886	60486
221 137		**XC**	HX	*XC*	CZ	60387	60787	60987	60887	60487

▲ The London Overground Class 172s entered traffic in summer 2010. On 24/07/10 172 005 leaves Upper Holloway with the 16.09 Barking–Gospel Oak.
Robert Pritchard

▼ Arriva Trains-liveried 175 010 leaves Crewe with the 18.30 Manchester Piccadilly–Carmarthen on 09/08/10.
Robert Pritchard

▲ TransPennine Express Desiro 185 106 passes Docker on the WCML with the 12.08 Edinburgh–Manchester Airport on 21/04/10. **Peter Foster**

▼ First Hull Trains-liveried 180 113 passes Yaxley, just south of Peterborough, with the 06.25 Hull–London King's Cross on 08/04/10. **Peter Foster**

▲ CrossCountry-liveried 220 007 is seen on the approaches to Chesterfield with the 06.21 Newcastle–Reading on 04/06/10. **Robert Pritchard**

▲ Virgin Trains-liveried 221 101 passes Crawford, on the climb to Beattock Summit, with the 14.52 Edinburgh–Birmingham New Street on 11/09/10.
Robin Ralston

▼ East Midlands Trains Meridians 222 013 and 222 015 pass Kibworth, between Leicester and Market Harborough, with the 11.47 Sheffield–London St Pancras on 20/05/10.
Lindsay Atkinson

▲ Balfour Beatty Rail Services Plasser & Theurer 07-16 Universal Tamper DR 73263 at Yeovil Junction on 04/03/10. **Stacey Thew**

▼ Plasser & Theurer 08-16/4x4C100-RT Tamper DR 73919 at Barrow Hill Roundhouse after having been repainted in the livery of its operator, Colas Rail, on 22/08/10. **Mick Tindall**

▲ VolkerRail Matisa B41 UE Tampers DR 75405 and DR 75404 pass Waitby on the Settle & Carlisle line on 16/07/10. **Jamie Squibbs**

▼ Network Rail Pandrol Jackson Plain Line Stoneblower DR 80206 passes Coaley, Gloucestershire on 02/07/10. **Jamie Squibbs**

▲ Colas Rail Geismar GP-TRAMM VMT 860 PL/UM with trailer DR 98308A+
DR 98308B stabled at Gloucester on 05/12/09. **Stacey Thew**

▼ Network Rail Windhoff Multi Purpose Vehicle Master & Slave DR 98961+
DR 98911 is seen at Inverness on 23/05/10. These vehicles are used across the
country on weedkilling, Sandite and de-icing duties. **Alexander Colley**

▲ One of three Loram Rail/Network Rail C21 Rail Grinding Trains, set DR 79257+ DR 79256+DR 79255+DR 79254+DR 79253+DR 79252+DR 79251 (nearest camera) is seen heading north towards Rotherham near Beighton Junction on 08/08/09. **Robert Pritchard**

▼ Network Rail Plasser & Theurer EM-SAT 100/RT Track Survey Car 999801 heads south on the WCML at Tamworth on 12/07/10. **Jamie Squibbs**

221 138		**XC**	HX	*XC*	CZ	60388	60788	60988	60888	60488
221 139		**XC**	HX	*XC*	CZ	60389	60789	60989	60889	60489
221 140		**XC**	HX	*XC*	CZ	60390	60790	60990	60890	60490
221 141		**XC**	HX	*XC*	CZ	60391	60791	60991		60491
221 142	*	**VT**	HX	*VW*	CZ	60392	60992		60792	60492
221 143	*	**VT**	HX	*VW*	CZ	60393	60993		60793	60493
221 144	*	**VT**	HX	*VW*	CZ	60394	60994		60794	60494

Names (carried on MS No. 609xx):

221 101	Louis Bleriot		221 110	James Cook
221 102	John Cabot		221 111	Roald Amundsen
221 103	Christopher Columbus		221 112	Ferdinand Magellan
221 104	Sir John Franklin		221 113	Sir Walter Raleigh
221 105	William Baffin		221 115	Polmadie Depot
221 106	Willem Barents		221 142	Matthew Flinders
221 107	Sir Martin Frobisher		221 143	Auguste Picard
221 108	Sir Ernest Shackleton		221 144	BOMBARDIER Voyager
221 109	Marco Polo			

CLASS 222   MERIDIAN   BOMBARDIER

Construction: Steel.
Engine: Cummins QSK19 of 560 kW (750 h.p.) at 1800 r.p.m.
Transmission: Two Alstom Onix 800 three-phase traction motors of 275 kW.
Braking: Rheostatic and electro-pneumatic.
Bogies: Bombardier B5005.
Couplers: Dellner at outer ends, bar within unit.
Dimensions: 23.85/23.00 x 2.73 m.
Gangways: Within unit only.   **Wheel Arrangement:** All cars 1A-A1.
Doors: Single-leaf swing plug.   **Maximum Speed:** 125 m.p.h.
Seating Layout: 1: 2+1, 2: 2+2 facing/unidirectional.
Multiple Working: Within class and with Classes 220 and 221 (in an emergency).

222 001–222 006. 7-car units. DMF–MF–MF–MSRMB–MS–MS–DMS.

Note: The 7-car units were built as 9-car units, before being reduced to 8-car sets and then later to 7-car sets to strengthen all 4-car units to 5-cars. 222 007 was built as a 9-car unit but has now been reduced to a 5-car unit.

DMF. Bombardier Brugge 2004–2005. 22/– 1TD 1W. 52.8 t.
MF. Bombardier Brugge 2004–2005. 42/– 1T. 46.8 t.
MSRMB. Bombardier Brugge 2004–2005. –/62. 48.0 t.
MS. Bombardier Brugge 2004–2005. –/68 1T. 47.0 t.
DMS. Bombardier Brugge 2004–2005. –/38 1TD 1W. 49.4 t.

222 001	**ST**	E	*EM*	DY	60241	60445	60341	60621
					60561	60551	60161	
222 002	**ST**	E	*EM*	DY	60242	60446	60342	60622
					60562	60544	60162	
222 003	**ST**	E	*EM*	DY	60243	60446	60343	60623
					60563	60553	60163	TORNADO

222 004	**ST**	E	*EM*	DY	60244	60345	60344	60624	
					60564	60554	60164		
222 005	**ST**	E	*EM*	DY	60245	60347	60443	60625	
					60565	60555	60165		
222 006	**ST**	E	*EM*	DY	60246	60447	60441	60626	
					60566	60556	60166		

222 007–222 023. 5-car units. DMF–MC–MSRMB–MS–DMS.

DMF. Bombardier Brugge 2003–2004. 22/– 1TD 1W. 52.8 t.
MC. Bombardier Brugge 2003–2004. 28/22 1T. 48.6 t.
MSRMB. Bombardier Brugge 2003–2004. –/62. 49.6 t.
MS. Bombardier Brugge 2004–2005. –/68 1T. 47.0 t.
DMS. Bombardier Brugge 2003–2004. –/40 1TD 1W. 51.0 t.

222 007	**ST**	E	*EM*	DY	60247	60442	60627	60567	60167
222 008	**ST**	E	*EM*	DY	60248	60918	60628	60545	60168
222 009	**ST**	E	*EM*	DY	60249	60919	60629	60557	60169
222 010	**ST**	E	*EM*	DY	60250	60920	60630	60546	60170
222 011	**ST**	E	*EM*	DY	60251	60921	60631	60531	60171
222 012	**ST**	E	*EM*	DY	60252	60922	60632	60532	60172
222 013	**ST**	E	*EM*	DY	60253	60923	60633	60533	60173
222 014	**ST**	E	*EM*	DY	60254	60924	60634	60534	60174
222 015	**ST**	E	*EM*	DY	60255	60925	60635	60535	60175
222 016	**ST**	E	*EM*	DY	60256	60926	60636	60536	60176
222 017	**ST**	E	*EM*	DY	60257	60927	60637	60537	60177
222 018	**ST**	E	*EM*	DY	60258	60928	60638	60444	60178
222 019	**ST**	E	*EM*	DY	60259	60929	60639	60547	60179
222 020	**ST**	E	*EM*	DY	60260	60930	60640	60543	60180
222 021	**ST**	E	*EM*	DY	60261	60931	60641	60552	60181
222 022	**ST**	E	*EM*	DY	60262	60932	60642	60542	60182
222 023	**ST**	E	*EM*	DY	60263	60933	60643	60541	60183

222 101–222 104. 4-car units formerly operated by Hull Trains. DMF–MC–MSRMB–DMS.

DMF. Bombardier Brugge 2005. 22/– 1TD 1W. 52.8 t.
MC. Bombardier Brugge 2005. 11/46 1T. 47.1 t.
MSRMB. Bombardier Brugge 2005. –/62. 48.0 t.
DMS. Bombardier Brugge 2005. –/40 1TD 1W. 49.4 t.

222 101	**ST**	E	*EM*	DY	60271	60571	60681	60191
222 102	**ST**	E	*EM*	DY	60272	60572	60682	60192
222 103	**ST**	E	*EM*	DY	60273	60573	60683	60193
222 104	**ST**	E	*EM*	DY	60274	60574	60684	60194

3. SERVICE DMUS

This section lists vehicles not used for passenger-carrying purposes. Vehicles are numbered in the special service stock number series.

CLASS 950 — TRACK ASSESSMENT UNIT

DM–DM. Purpose built service unit based on the Class 150/1 design. Gangwayed within unit.

Construction: Steel.
Engine: One Cummins NT-855-RT5 of 213 kW (285 h.p.) at 2100 r.p.m. per power car.
Transmission: Hydraulic. Voith T211r with cardan shafts to Gmeinder GM190 final drive.
Maximum Speed: 75 m.p.h.   **Couplers:** BSI automatic.
Bogies: BP38 (powered), BT38 (non-powered).
Brakes: Electro-pneumatic.   **Dimensions:** 20.06 x 2.82 m.
Doors: Manually operated slam & power operated sliding.
Multiple Working: Classes 142, 143, 144, 150, 153, 155, 156, 158, 159 and 170.

999600. DM. Lot No. 4060 BREL York 1987. 35.0 t.
999601. DM. Lot No. 4061 BREL York 1987. 35.0 t.

950 001   **Y**   NR *DB*   ZA   999600 999601

CLASS 960 — SANDITE & SERVICE UNITS

DMB. Converted from Class 121. Non gangwayed.
For details see Page 9.
This is a Chiltern Route Learning Unit and is also hired to other operators.
Lot No. 30518 Pressed Steel 1960. 38.0 t.

960 014   **BG**   CR *CR*   AL   977873   (55022)

CLASS 960   WATER-JETTING UNIT

DMB–MS–DMB. Converted 2003–2004 from Class 117. Non gangwayed.

Construction: Steel.
Engines: Two Leyland 1595 of 112 kW (150 h.p.) at 1800 r.p.m.
Transmission: Mechanical. Cardan shaft and freewheel to a four-speed
epicyclic gearbox with a further cardan shaft to the final drive, each engine
driving the inner axle of one bogie.
Maximum Speed: 70 m.p.h.
Bogies: DD10.   **Couplings:** Screw.
Brakes: Twin pipe vacuum.   **Multiple Working:** Blue Square.
Doors: Manually operated slam.   **Dimensions:** 20.45 x 2.84 m.

977987/977988. DMB. Lot No. 30546/30548 Pressed Steel 1959–1960. 36.5 t.
977992. MS. Lot No. 30548 Pressed Steel 1959–1960. 36.5 t.

960 301	**G**	CR	*CR*	AL	977987	(51371)	977992 (51375)
					977988	(51413)	

4. DMUS AWAITING DISPOSAL

The list below comprises vehicles awaiting disposal which are stored on the
National Railway network.

Class 101

Spare	**RR**	X	SN	51432	51498

Class 960

Converted from Class 121 or 122. 960 302/303 converted for use as Severn
Tunnel Emergency Train units, but not actually used as such.

960 010	**M**	NR	AL	977858	(55024)
960 011	**RK**	NR	TM	977859	(55025)
960 013	**N**	NR	AL	977866	(55030)
960 015	**Y**	NR	AL	975042	(55019)
960 021	**R0**	NR	AL	977723	(55021)
-	**Y**	CS	RU	977968	(55029)
960 302	**Y**	AW	CF	977975	(55027)
960 303	**Y**	AW	CF	977976	(55031)

5. ON-TRACK MACHINES

These machines are used for maintaining, renewing and enhancing the infrastructure of the national railway network. With the exception of snowploughs all can be self-propelled. They are permitted to operate either under their own power or in train formations throughout the network both within and outside engineering possessions. Machines only permitted to be used within engineering possessions, referred to as On-Track Plant, are not included. For each machine its Network Rail registered number, owner/responsible custodian and type is given, plus its name if carried. Actual operation of each machine is undertaken by either the owner/responsible custodian or a contracted responsible custodian.

DYNAMIC TRACK STABILISERS

DR 72201	FA	Plasser & Theurer DGS 62-N
DR 72203	FA	Plasser & Theurer DGS 62-N
DR 72205	FA	Plasser & Theurer DGS 62-N
DR 72206	FA	Plasser & Theurer DGS 62-N
DR 72208	FA	Plasser & Theurer DGS 62-N
DR 72209	FA	Plasser & Theurer DGS 62-N
DR 72211	BB	Plasser & Theurer DGS 62-N
DR 72213	BB	Plasser & Theurer DGS 62-N

TAMPERS

DR 73101	FA	Plasser & Theurer 09-32 CSM	
DR 73103	CS	Plasser & Theurer 09-32 CSM	
DR 73104	FA	Plasser & Theurer 09-32 CSM	
DR 73105	CS	Plasser & Theurer 09-32 CSM	
DR 73106	FA	Plasser & Theurer 09-32 CSM	
DR 73107	FA	Plasser & Theurer 09-32 CSM	
DR 73108	AY	Plasser & Theurer 09-32-RT	Tiger
DR 73109	SK	Plasser & Theurer 09-3X-RT	
DR 73110	SK	Plasser & Theurer 09-3X-RT	Peter White
DR 73111	NR	Plasser & Theurer 09-3X-Dynamic	Reading Panel 1965–2005
DR 73113	NR	Plasser & Theurer 09-3X-Dynamic	
DR 73114	NR	Plasser & Theurer 09-3X-Dynamic	Ron Henderson
DR 73115	NR	Plasser & Theurer 09-3X-Dynamic	
DR 73116	NR	Plasser & Theurer 09-3X-Dynamic	
DR 73117	NR	Plasser & Theurer 09-3X Dynamic	
DR 73118	NR	Plasser & Theurer 09-3X Dynamic	
DR 73238	FA	Plasser & Theurer 07-16 Universal	Brian Langley
DR 73243	FA	Plasser & Theurer 07-16 Universal	
DR 73244	FA	Plasser & Theurer 07-16 Universal	
DR 73246	BB	Plasser & Theurer 07-16 Universal	
DR 73248	FA	Plasser & Theurer 07-16 Universal	
DR 73251	BB	Plasser & Theurer 07-16 Universal	
DR 73256	FA	Plasser & Theurer 07-16 Universal	
DR 73257	BB	Plasser & Theurer 07-16 Universal	

DR 73261	BB	Plasser & Theurer 07-16 Universal	
DR 73263	BB	Plasser & Theurer 07-16 Universal	
DR 73265	FA	Plasser & Theurer 07-16 Universal	
DR 73266	BB	Plasser & Theurer 07-16 Universal	
DR 73267	FA	Plasser & Theurer 07-16 Universal	
DR 73268	FA	Plasser & Theurer 07-16 Universal	
DR 73269	FA	Plasser & Theurer 07-16 Universal	
DR 73270	FA	Plasser & Theurer 07-16 Universal	Alan Chamberlain
DR 73271	FA	Plasser & Theurer 07-16 Universal	
DR 73272	FA	Plasser & Theurer 07-16 Universal	
DR 73273	BB	Plasser & Theurer 07-16 Universal	
DR 73274	FA	Plasser & Theurer 07-16 Universal	
DR 73276	BB	Plasser & Theurer 07-16 Universal	
DR 73278	BB	Plasser & Theurer 07-16 Universal	
DR 73307	FA	Plasser & Theurer 07-275 Switch & Crossing	
DR 73309	FA	Plasser & Theurer 07-275 Switch & Crossing	
DR 73310	FA	Plasser & Theurer 07-275 Switch & Crossing	
DR 73311	BB	Plasser & Theurer 07-275 Switch & Crossing	Cyril Dryland
DR 73312	FA	Plasser & Theurer 07-275 Switch & Crossing	
DR 73314	FA	Plasser & Theurer 07-275 Switch & Crossing	
DR 73316	FA	Plasser & Theurer 07-275 Switch & Crossing	
DR 73318	BB	Plasser & Theurer 07-275 Switch & Crossing	Peter Atkinson
DR 73321	FA	Plasser & Theurer 07-275 Switch & Crossing	
DR 73403	FA	Plasser & Theurer 07-32 Duomatic	
DR 73404	FA	Plasser & Theurer 07-32 Duomatic	
DR 73413	FA	Plasser & Theurer 07-32 Duomatic	
DR 73414	FA	Plasser & Theurer 07-32 Duomatic	
DR 73415	FA	Plasser & Theurer 07-32 Duomatic	
DR 73416	FA	Plasser & Theurer 07-32 Duomatic	
DR 73418	FA	Plasser & Theurer 07-32 Duomatic	
DR 73419	FA	Plasser & Theurer 07-32 Duomatic	
DR 73420	FA	Plasser & Theurer 07-32 Duomatic	
DR 73421	FA	Plasser & Theurer 07-32 Duomatic	
DR 73423	FA	Plasser & Theurer 07-32 Duomatic	
DR 73424	BB	Plasser & Theurer 07-32 Duomatic	
DR 73425	FA	Plasser & Theurer 07-32 Duomatic	
DR 73426	FA	Plasser & Theurer 07-32 Duomatic	
DR 73427	FA	Plasser & Theurer 07-32 Duomatic	
DR 73428	FA	Plasser & Theurer 07-32 Duomatic	
DR 73431	FA	Plasser & Theurer 07-32 Duomatic	
DR 73432	FA	Plasser & Theurer 07-32 Duomatic	
DR 73433	FA	Plasser & Theurer 07-32 Duomatic	
DR 73434	BB	Plasser & Theurer 07-32 Duomatic	
DR 73435	FA	Plasser & Theurer 07-32 Duomatic	
DR 73502	BB	Plasser & Theurer 08-16/90 ZW	
DR 73503	BB	Plasser & Theurer 08-16/90 ZW	
DR 73601	FA	Plasser & Theurer 07-32 Duomatic	
DR 73802	FA	Plasser & Theurer 08-16 Universal	
DR 73803	SK	Plasser & Theurer 08-32U-RT	Alexander Graham Bell
DR 73804	SK	Plasser & Theurer 08-32U-RT	James Watt
DR 73805	CS	Plasser & Theurer 08-16(32)U-RT	

DR 73806	CS	Plasser & Theurer 08-16/32U-RT	Karine
DR 73901	CS	Plasser & Theurer 08-275 Switch & Crossing	
DR 73902	FA	Plasser & Theurer 08-275 Switch & Crossing	
DR 73903	FA	Plasser & Theurer 08-275 Switch & Crossing	George Mullineux
DR 73904	SK	Plasser & Theurer 08-4x4/4S-RT	Thomas Telford
DR 73905	AY	Plasser & Theurer 08-4x4/4S-RT	Eddie King
DR 73906	AY	Plasser & Theurer 08-4x4/4S-RT	Panther
DR 73907	CS	Plasser & Theurer 08-4x4/4S-RT	
DR 73908	CS	Plasser & Theurer 08-4x4/4S-RT	
DR 73909	CS	Plasser & Theurer 08-4x4/4S-RT	Saturn
DR 73910	CS	Plasser & Theurer 08-4x4/4S-RT	Jupiter
DR 73911	AY	Plasser & Theurer 08-16/4x4C-RT	Puma
DR 73912	AY	Plasser & Theurer 08-16/4x4C-RT	Lynx
DR 73913	CS	Plasser & Theurer 08-12/4x4C-RT	
DR 73914	SK	Plasser & Theurer 08-4x4/4S-RT	Robert McAlpine
DR 73915	SK	Plasser & Theurer 08-16/4x4C-RT	William Arrol
DR 73916	SK	Plasser & Theurer 08-16/4x4C-RT	First Engineering
DR 73917	BB	Plasser & Theurer 08-4x4/4S-RT	
DR 73918	BB	Plasser & Theurer 08-4x4/4S-RT	
DR 73919	CS	Plasser & Theurer 08-16/4x4C100-RT	
DR 73920	AY	Plasser & Theurer 08-16/4x4C80-RT	
DR 73921	AY	Plasser & Theurer 08-16/4x4C80-RT	John Snowdon
DR 73922	AY	Plasser & Theurer 08-16/4x4C80-RT	
DR 73923	CS	Plasser & Theurer 08-4x4/4S-RT	Mercury
DR 73924	CS	Plasser & Theurer 08-16/4x4C100-RT	Atlas
DR 73925	CS	Plasser & Theurer 08-16/4x4C100-RT	Europa
DR 73926	BB	Plasser & Theurer 08-16/4x4C100-RT	Stephen Keith Blanchard
DR 73927	BB	Plasser & Theurer 08-16/4x4C100-RT	
DR 73928	BB	Plasser & Theurer 08-16/4x4C100-RT	
DR 73929	CS	Plasser & Theurer 08-4x4/4S-RT	
DR 73930	CS	Plasser & Theurer 08-4x4/4S-RT	
DR 73931	CS	Plasser & Theurer 08-16/4x4C100-RT	
DR 73932	SK	Plasser & Theurer 08-4x4/4S-RT	
DR 73933	SK	Plasser & Theurer 08-16/4x4/C100-RT	
DR 73934	SK	Plasser & Theurer 08-16/4x4/C100-RT	
DR 73935	CS	Plasser & Theurer 08-4x4/4S-RT	
DR 73936	CS	Plasser & Theurer 08-4x4/4S-RT	
DR 73937	BB	Plasser & Theurer 08-16/4x4C100-RT	
DR 73938	BB	Plasser & Theurer 08-16/4x4C100-RT	
DR 73939	BB	Plasser & Theurer 08-16/4x4C100-RT	
DR 73940	SK	Plasser & Theurer 08-4x4/4S-RT	
DR 73941	SK	Plasser & Theurer 08-4x4/4S-RT	
DR 73942	CS	Plasser & Theurer 08-4x4/4S-RT	
DR 73943	BB	Plasser & Theurer 08-16/4x4C100-RT	
DR 73944	BB	Plasser & Theurer 08-16/4x4C100-RT	
DR 73945	BB	Plasser & Theurer 08-16/4x4C100-RT	
DR 73946	VO	Plasser & Theurer Euromat 08-4x4/4S	
DR 73947	CS	Plasser & Theurer 08-4x4/4S-RT	
DR 73948	CS	Plasser & Theurer 08-4x4/4S-RT	
DR 75001	FA	Plasser & Theurer 08-16/90	
DR 75201	BB	Plasser & Theurer 08-275 Switch & Crossing	

DR 75202	BB	Plasser & Theurer 08-275 Switch & Crossing
DR 75203	FA	Plasser & Theurer 08-75 SP-T
DR 75301	VO	Matisa B 45 UE
DR 75302	VO	Matisa B 45 UE
DR 75303	VO	Matisa B 45 UE
DR 75401	VO	Matisa B 41 UE
DR 75402	VO	Matisa B 41 UE
DR 75403	VO	Matisa B 41 UE
DR 75404	VO	Matisa B 41 UE
DR 75405	VO	Matisa B 41 UE
DR 75406	CS	Matisa B41 UE
DR 75407	CS	Matisa B41 UE
DR 75408	BB	Matisa B41 UE
DR 75409	BB	Matisa B41 UE
DR 75410	BB	Matisa B41 UE
DR 75411	BB	Matisa B41 UE
DR 75501	BB	Matisa B66 UC
DR 75502	BB	Matisa B66 UC
DR 76801	NR	Plasser & Theurer 09-CM-NR

BALLAST CLEANERS

DR 76304	FA	Plasser & Theurer RM74	
DR 76311	FA	Plasser & Theurer RM74	
DR 76318	FA	Plasser & Theurer RM74	
DR 76323	NR	Plasser & Theurer RM95-RT	
DR 76324	NR	Plasser & Theurer RM95-RT	
DR 76501	NR	Plasser & Theurer RM-900-RT	
DR 76502	NR	Plasser & Theurer RM-900-RT	
DR 76503	NR	Plasser & Theurer RM-900-RT	
DR 76601	CS	Plasser & Theurer RM90-NR	Olwen
DR 76701	NR	Plasser & Theurer VM80-NR	
DR 76710	NR	Plasser & Theurer VM80-NR	
DR 76750	NR	Matisa D75	

FINISHING MACHINES AND REGULATORS

DR 77001	SK	Plasser & Theurer AFM 2000-RT Finishing Machine
DR 77002	SK	Plasser & Theurer AFM 2000-RT Finishing Machine
DR 77313	FA	Plasser & Theurer USP 5000C Regulator
DR 77315	BB	Plasser & Theurer USP 5000C Regulator
DR 77316	BB	Plasser & Theurer USP 5000C Regulator
DR 77317	FA	Plasser & Theurer USP 5000C Regulator
DR 77319	CS	Plasser & Theurer USP 5000C Regulator
DR 77320	FA	Plasser & Theurer USP 5000C Regulator
DR 77321	FA	Plasser & Theurer USP 5000C Regulator
DR 77322	BB	Plasser & Theurer USP 5000C Regulator
DR 77323	FA	Plasser & Theurer USP 5000C Regulator
DR 77325	FA	Plasser & Theurer USP 5000C Regulator
DR 77326	FA	Plasser & Theurer USP 5000C Regulator
DR 77327	CS	Plasser & Theurer USP 5000C Regulator

)R 77328	FA	Plasser & Theurer USP 5000C Regulator	
)R 77329	FA	Plasser & Theurer USP 5000C Regulator	
)R 77330	FA	Plasser & Theurer USP 5000C Regulator	
)R 77331	FA	Plasser & Theurer USP 5000C Regulator	
)R 77332	FA	Plasser & Theurer USP 5000C Regulator	
)R 77333	FA	Plasser & Theurer USP 5000C Regulator	
)R 77335	CS	Plasser & Theurer USP 5000C Regulator	
)R 77336	BB	Plasser & Theurer USP 5000C Regulator	
)R 77801	VO	Matisa R 24 S Regulator	
)R 77802	VO	Matisa R 24 S Regulator	
)R 77901	CS	Plasser & Theurer USP 5000-RT Regulator	
)R 77903	NR	Plasser & Theurer USP 5000-RT Regulator	Frank Jones
)R 77904	NR	Plasser & Theurer USP 5000-RT Regulator	
)R 77905	NR	Plasser & Theurer USP 5000-RT Regulator	
)R 77906	NR	Plasser & Theurer USP 5000-RT Regulator	
)R 77907	NR	Plasser & Theurer USP 5000-RT Regulator	
)R 77908	SK	Plasser & Theurer USP 5000-RT Regulator	

TWIN JIB TRACK RELAYERS

)RS 78211	FA	Plasser & Theurer Self-Propelled Heavy Duty
)RP 78212	FA	Plasser & Theurer Self-Propelled Heavy Duty
)RP 78213	VO	Plasser & Theurer Self-Propelled Heavy Duty
)RP 78214	FA	Plasser & Theurer Self-Propelled Heavy Duty
)RP 78215	FA	Plasser & Theurer Self-Propelled Heavy Duty
)RP 78216	BB	Plasser & Theurer Self-Propelled Heavy Duty
)RP 78217	FA	Plasser & Theurer Self-Propelled Heavy Duty
)RP 78218	BB	Plasser & Theurer Self-Propelled Heavy Duty
)RP 78219	FA	Plasser & Theurer Self-Propelled Heavy Duty
)RP 78221	BB	Plasser & Theurer Self-Propelled Heavy Duty
)RP 78222	BB	Plasser & Theurer Self-Propelled Heavy Duty
)RP 78223	BB	Plasser & Theurer Self-Propelled Heavy Duty
)RP 78224	BB	Plasser & Theurer Self-Propelled Heavy Duty
)RC 78225	FA	Cowans Sheldon Self-Propelled Heavy Duty
)RC 78226	FA	Cowans Sheldon Self-Propelled Heavy Duty
)RC 78227	FA	Cowans Sheldon Self-Propelled Heavy Duty
)RC 78229	FA	Cowans Sheldon Self-Propelled Heavy Duty
)RC 78230	FA	Cowans Sheldon Self-Propelled Heavy Duty
)RC 78231	FA	Cowans Sheldon Self-Propelled Heavy Duty
)RC 78232	FA	Cowans Sheldon Self-Propelled Heavy Duty
)RC 78233	FA	Cowans Sheldon Self-Propelled Heavy Duty
)RC 78234	FA	Cowans Sheldon Self-Propelled Heavy Duty
)RC 78235	FA	Cowans Sheldon Self-Propelled Heavy Duty
)RC 78237	FA	Cowans Sheldon Self-Propelled Heavy Duty

TRACK RENEWAL MACHINES

Matisa P95 Track Renewals Trains

|)R 78801 | + | DR 78811 | + | DR 78821 | + | DR 78831 | NR |
|)R 78802 | + | DR 78812 | + | DR 78822 | + | DR 78832 | NR |

RAIL GRINDING TRAINS

Loram SPML 15
DR 79200A + DR 79200B + DR 79200C   LO

Loram SPML 17
DR 79201A + DR 79201B   LO

Speno RPS-32
DR 79221 + DR 79222 + DR 79223 + DR 79224 + DR 79225 + DR 79226   SI

Loram C21
DR 79231 + DR 79232 + DR 79233 + DR 79234 + DR 79235 + DR 79236 + DR 79237   LO
DR 79241 + DR 79242 + DR 79243 + DR 79244 + DR 79245 + DR 79246 + DR 79247   NR
DR 79251 + DR 79252 + DR 79253 + DR 79254 + DR 79255 + DR 79256 + DR 79257   NR

Harsco Track Technologies RGH Switch & Crossing 20C

DR 79261 + DR 79271	NR	
DR 79262 + DR 79272	NR	
DR 79263 + DR 79273	NR	
DR 79264 + DR 79274	NR	
DR 79265	NR	*spare vehicle*

STONE BLOWERS

DR 80200	NR	Pandrol Jackson Plain Line
DR 80201	NR	Pandrol Jackson Plain Line
DR 80202	NR	Pandrol Jackson Plain Line
DR 80203	NR	Pandrol Jackson Plain Line
DR 80204	NR	Pandrol Jackson Plain Line
DR 80205	NR	Pandrol Jackson Plain Line
DR 80206	NR	Pandrol Jackson Plain Line
DR 80207	NR	Pandrol Jackson Plain Line
DR 80208	NR	Pandrol Jackson Plain Line
DR 80209	NR	Pandrol Jackson Plain Line
DR 80210	NR	Pandrol Jackson Plain Line
DR 80211	NR	Pandrol Jackson Plain Line
DR 80212	NR	Pandrol Jackson Plain Line
DR 80213	NR	Harsco Track Technologies Plain Line
DR 80214	NR	Harsco Track Technologies Plain Line
DR 80215	NR	Harsco Track Technologies Plain Line
DR 80216	NR	Harsco Track Technologies Plain Line
DR 80217	NR	Harsco Track Technologies Plain Line
DR 80301	NR	Harsco Track Technologies Multi-purpose Stephen Cornis
DR 80302	NR	Harsco Track Technologies Multi-purpose
DR 80303	NR	Harsco Track Technologies Multi-purpose

CRANES

DRP 81505	BB	Plasser & Theurer 12 tonne Heavy Duty Diesel Hydraulic
DRP 81507	BB	Plasser & Theurer 12 tonne Heavy Duty Diesel Hydraulic
DRP 81508	BB	Plasser & Theurer 12 tonne Heavy Duty Diesel Hydraulic
DRP 81511	BB	Plasser & Theurer 12 tonneHeavy Duty Diesel Hydraulic

DRP 81513	BB	Plasser & Theurer 12 tonne Heavy Duty Diesel Hydraulic
DRP 81514	FA	Plasser & Theurer 12 tonne Heavy Duty Diesel Hydraulic
DRP 81515	FA	Plasser & Theurer 12 tonne Heavy Duty Diesel Hydraulic
DRP 81517	BB	Plasser & Theurer 12 tonne Heavy Duty Diesel Hydraulic
DRP 81519	BB	Plasser & Theurer 12 tonne Heavy Duty Diesel Hydraulic
DRP 81521	FA	Plasser & Theurer 12 tonne Heavy Duty Diesel Hydraulic
DRP 81522	BB	Plasser & Theurer 12 tonne Heavy Duty Diesel Hydraulic
DRP 81525	BB	Plasser & Theurer 12 tonne Heavy Duty Diesel Hydraulic
DRP 81527	FA	Plasser & Theurer 12 tonne Heavy Duty Diesel Hydraulic
DRP 81528	FA	Plasser & Theurer 12 tonne Heavy Duty Diesel Hydraulic
DRP 81529	FA	Plasser & Theurer 12 tonne Heavy Duty Diesel Hydraulic
DRP 81532	BB	Plasser & Theurer 12 tonne Heavy Duty Diesel Hydraulic
DRK 81601	VO	Kirow KRC810UK 100 tonne Heavy Duty Diesel Hydraulic
DRK 81602	BB	Kirow KRC810UK 100 tonne Heavy Duty Diesel Hydraulic
DRK 81611	BB	Kirow KRC1200UK 125 tonne Heavy Duty Diesel Hydraulic
DRK 81612	CS	Kirow KRC1200UK 125 tonne Heavy Duty Diesel Hydraulic
DRK 81613	VO	Kirow KRC1200UK 125 tonne Heavy Duty Diesel Hydraulic
DRK 81621	VO	Kirow KRC250UK 25 tonne Diesel Hydraulic
DRK 81622	VO	Kirow KRC250UK 25 tonne Diesel Hydraulic
DRK 81623	SK	Kirow KRC250UK 25 tonne Diesel Hydraulic
DRK 81624	SK	Kirow KRC250UK 25 tonne Diesel Hydraulic
DRK 81625	SK	Kirow KRC250UK 25 tonne Diesel Hydraulic
ADRC 96710	NR	Cowans Sheldon 75 tonne Diesel Hydraulic (telescopic) Breakdown
ADRC 96713	NR	Cowans Sheldon 75 tonne Diesel Hydraulic (telescopic) Breakdown
ADRC 96714	NR	Cowans Sheldon 75 tonne Diesel Hydraulic (telescopic) Breakdown
ADRC 96715	NR	Cowans Sheldon 75 tonne Diesel Hydraulic (telescopic) Breakdown

Names: 81601   Nigel Chester   81611   Malcolm L. Pearce

GENERAL PURPOSE MAINTENANCE VEHICLES

DR 97001	H1	Eiv de Brieve DU94BA TRAMM with Crane	
DR 97011	H1	Windhoff MPV Master	
DR 97012	H1	Windhoff MPV Master	Geoff Bell
DR 97013	H1	Windhoff MPV Master	
DR 97014	H1	Windhoff MPV Master	
DR 98001	NR	Windhoff MPV Master with Piling Equipment	
DR 98002	NR	Windhoff MPV Master with Piling Equipment	
DR 98003	NR	Windhoff MPV Master with Overhead Line Renewal Equipment	
DR 98004	NR	Windhoff MPV Master with Overhead Line Renewal Equipment	
DR 98005	NR	Windhoff MPV Master with Piling Equipment	
DR 98006	NR	Windhoff MPV Master with Piling Equipment	
DR 98007	NR	Windhoff MPV Master with Piling Equipment	
DR 98008	NR	Windhoff MPV Twin-cab Master with GSM-R test equipment	
DR 98009	NR	Windhoff MPV Master with Overhead Line Renewal Equipment	
DR 98010	NR	Windhoff MPV Master with Overhead Line Renewal Equipment	
DR 98011	NR	Windhoff MPV Master with Overhead Line Renewal Equipment	
DR 98012	NR	Windhoff MPV Master with Overhead Line Renewal Equipment	
DR 98013	NR	Windhoff MPV Master with Overhead Line Renewal Equipment	
DR 98014	NR	Windhoff MPV Master with Overhead Line Renewal Equipment	
DR 98210A + DR 98210B	AY	Plasser & Theurer GP-TRAMM with Trailer	
DR 98215A + DR 98215B	BB	Plasser & Theurer GP-TRAMM with Trailer	

DR 98216A + DR 98216B	BB	Plasser & Theurer GP-TRAMM with Trailer
DR 98217A + DR 98217B	BB	Plasser & Theurer GP-TRAMM with Trailer
DR 98218A + DR 98218B	BB	Plasser & Theurer GP-TRAMM with Trailer
DR 98219A + DR 98219B	BB	Plasser & Theurer GP-TRAMM with Trailer
DR 98220A + DR 98220B	BB	Plasser & Theurer GP-TRAMM with Trailer
DR 98305	NR	Geismar GP-TRAMM VMT 860 PL/UM
DR 98306	NR	Geismar GP-TRAMM VMT 860 PL/UM
DR 98307A + DR 98307B	CS	Geismar GP-TRAMM VMT 860 PL/UM with Trailer
DR 98308A + DR 98308B	CS	Geismar GP-TRAMM VMT 860 PL/UM with Trailer
DR 98901 + DR 98951	NR	Windhoff MPV Master & Slave
DR 98902 + DR 98952	NR	Windhoff MPV Master & Slave
DR 98903 + DR 98953	NR	Windhoff MPV Master & Slave
DR 98904 + DR 98954	NR	Windhoff MPV Master & Slave
DR 98905 + DR 98955	NR	Windhoff MPV Master & Slave
DR 98906 + DR 98956	NR	Windhoff MPV Master & Slave
DR 98907 + DR 98957	NR	Windhoff MPV Master & Slave
DR 98908 + DR 98958	NR	Windhoff MPV Master & Slave
DR 98909 + DR 98959	NR	Windhoff MPV Master & Slave
DR 98910 + DR 98960	NR	Windhoff MPV Master & Slave
DR 98911 + DR 98961	NR	Windhoff MPV Master & Slave
DR 98912 + DR 98962	NR	Windhoff MPV Master & Slave
DR 98913 + DR 98963	NR	Windhoff MPV Master & Slave
DR 98914 + DR 98964	NR	Windhoff MPV Master & Slave
DR 98915 + DR 98965	NR	Windhoff MPV Master & Slave
DR 98916 + DR 98966	NR	Windhoff MPV Master & Slave
DR 98917 + DR 98967	NR	Windhoff MPV Master & Slave
DR 98918 + DR 98968	NR	Windhoff MPV Master & Slave
DR 98919 + DR 98969	NR	Windhoff MPV Master & Slave
DR 98920 + DR 98970	NR	Windhoff MPV Master & Slave
DR 98921 + DR 98971	NR	Windhoff MPV Master & Slave
DR 98922 + DR 98972	NR	Windhoff MPV Master & Slave
DR 98923 + DR 98973	NR	Windhoff MPV Master & Slave
DR 98924 + DR 98974	NR	Windhoff MPV Master & Slave
DR 98925 + DR 98975	NR	Windhoff MPV Master & Slave
DR 98926 + DR 98976	NR	Windhoff MPV Master & Powered Slave
DR 98927 + DR 98977	NR	Windhoff MPV Master & Powered Slave
DR 98928 + DR 98978	NR	Windhoff MPV Master & Powered Slave
DR 98929 + DR 98979	NR	Windhoff MPV Master & Powered Slave
DR 98930 + DR 98980	NR	Windhoff MPV Master & Powered Slave
DR 98931 + DR 98981	NR	Windhoff MPV Master & Powered Slave
DR 98932 + DR 98982	NR	Windhoff MPV Master & Powered Slave

SNOWPLOUGHS

ADB 965203	NR	Independent Drift Plough
ADB 965206	NR	Independent Drift Plough
ADB 965208	NR	Independent Drift Plough
ADB 965209	NR	Independent Drift Plough
ADB 965210	NR	Independent Drift Plough
ADB 965211	NR	Independent Drift Plough

ADB 965217	NR	Independent Drift Plough
ADB 965219	NR	Independent Drift Plough
ADB 965223	NR	Independent Drift Plough
ADB 965224	NR	Independent Drift Plough
ADB 965230	NR	Independent Drift Plough
ADB 965231	NR	Independent Drift Plough
ADB 965232	NR	Independent Drift Plough
ADB 965233	NR	Independent Drift Plough
ADB 965234	NR	Independent Drift Plough
ADB 965235	NR	Independent Drift Plough
ADB 965236	NR	Independent Drift Plough
ADB 965237	NR	Independent Drift Plough
ADB 965240	NR	Independent Drift Plough
ADB 965241	NR	Independent Drift Plough
ADB 965242	NR	Independent Drift Plough
ADB 965243	NR	Independent Drift Plough
ADB 965576	NR	Beilhack Type PB600 Plough
ADB 965577	NR	Beilhack Type PB600 Plough
ADB 965578	NR	Beilhack Type PB600 Plough
ADB 965579	NR	Beilhack Type PB600 Plough
ADB 965580	NR	Beilhack Type PB600 Plough
ADB 965581	NR	Beilhack Type PB600 Plough
ADB 966096	NR	Beilhack Type PB600 Plough
ADB 966097	NR	Beilhack Type PB600 Plough
ADB 966098	NR	Beilhack Type PB600 Plough
ADB 966099	NR	Beilhack Type PB600 Plough

SNOWBLOWERS

ADB 968500	NR	Beilhack Self Propelled Rotary
ADB 968501	NR	Beilhack Self Propelled Rotary

INFRASTRUCTURE MONITORING VEHICLES

999800	NR	Plasser & Theurer EM-SAT 100/RT Track Survey Car
999801	NR	Plasser & Theurer EM-SAT 100/RT Track Survey Car

Name: 999800 Richard Spoors

ON-TRACK MACHINES AWAITING DISPOSAL

Tampers

DR 73216	Plasser & Theurer 07-16 Universal	Bristol Marsh Jn
DR 86101	Plasser & Theurer 08-16 Universal	Hither Green

Wet spot treatment machine

DR 76401	Plasser & Theurer	Rugby

Twin Jib track relayer

DRB 78123	British Hoist & Crane Non-Self-Propelled	Polmadie DHS

General purpose maintenance vehicles

DR 98300A+DR98300B	Geismar GP-TRAMM with Trailer	Hitchin
DR 98302A+DR98302B	Geismar GP-TRAMM with Trailer	Hitchin

6. CODES

6.1. LIVERY CODES

1	"One" (metallic grey with a broad black bodyside stripe. White National Express "interim" stripe as branding).
AL	Advertising/promotional livery (see class heading for details).
AN	Anglia Railways Class 170s (white & turquoise with blue vignette).
AR	Anglia Railways (turquoise blue with a white stripe).
AV	Arriva Trains (turquoise blue with white doors and a cream "swish").
BG	BR blue & grey lined out in white.
CI	Centro (light green with a broad blue lower bodyside band & blue cab end sections).
CR	Chiltern Railways (blue & white with a thin red stripe).
CT	Central Trains (two-tone green with yellow doors. Blue flash and red stripe at vehicle ends).
EM	East Midlands Trains {Connect} (blue with red & orange swish at unit ends)
FB	First Group dark blue.
FD	First Great Western & First Hull Trains "Dynamic Lines" (dark blue with thin multi-coloured lines on lower bodyside).
FG	First Group InterCity (indigo blue with a white roof & gold, pink & white stripes)
FI	First Great Western "Local Lines" DMU (varying blue with local visitor attractions applied to the lower bodyside).
FS	First Group (indigo blue with pink & white stripes).
FT	First TransPennine Express "Dynamic Lines" (varying blue with thin multi-coloured lines on lower bodyside).
G	BR Southern Region or BR DMU green.
GC	Grand Central (all over black with an orange stripe).
LM	London Midland (white/grey & green with broad black stripe around windows)
LO	London Overground (all over white with a blue solebar & black window surrounds).
M	BR maroon (maroon lined out in straw & black).
N	BR Network SouthEast (white & blue with red lower bodyside stripe, grey solebar & cab ends).
NO	Northern (deep blue, purple & white). Some units have area-specific promotional vinyls (see class headings for details).
NW	North Western Trains (blue with gold cantrail stripe & star).
NX	National Express (white with grey ends).
O	Non-standard livery (see class heading for details).
RK	Railtrack (green & blue).
RO	Old Railtrack (orange with white & grey stripes).
RR	Regional Railways (dark blue & grey with light blue & white stripes, three narrow dark blue stripes at vehicle ends).
SC	Strathclyde PTE (carmine & cream lined out in black & gold).
SL	Silverlink (indigo blue with white stripe, green lower body & yellow doors)
SN	Southern (white & dark green with light green semi-circles at one end of each vehicle. Light grey band at solebar level).
SP	Strathclyde PTE {revised} (carmine & cream, with a turquoise stripe)

SR ScotRail – Scotland's Railways (dark blue with Scottish Saltire flag &
white/light blue flashes).
ST Stagecoach {long-distance stock} (white & dark blue with dark blue
window surrounds and red & orange swishes at unit ends).
VT Virgin Trains silver (silver, with black window surrounds, white cantrail
stripe & red roof. Red swept down at unit ends. Black & white striped doors).
WB Wales & Borders Alphaline (metallic silver with blue doors).
WM Network West Midlands (light blue with green lower bodyside stripe
and white stripe at cantrail level).
WT Wessex Trains Alphaline (metallic silver with maroon or pink doors).
XC CrossCountry (two-tone silver with deep crimson ends & pink doors).
Y Network Rail yellow.

6.2. OWNER CODES

A	Angel Trains
AW	Arriva Trains Wales
AY	Amey Infrastructure Services
BB	Balfour Beatty Rail Infrastructure Services
BC	Bridgend County Borough Council/Rhondda Cynon Taff County Borough Council
CC	Cardiff City Council
CR	Chiltern Railways
CS	Colas Rail
E	Eversholt Rail (UK)
FA	Fastline (*in administration*)
H1	High Speed 1
HD	Hastings Diesels
HX	Halifax Bank of Scotland
LO	Loram
NR	Network Rail
P	Porterbrook Leasing Company
RI	Rail Assets Investments
SI	Speno International
SK	Swieteisky Babcock Rail
VO	VolkerRail
X	Sold for scrap/further use and awaiting collection or owner unknown

6.3. OPERATOR CODES

AW	Arriva Trains Wales
CR	Chiltern Railways
DB	DB Schenker Rail (UK)
EA	National Express East Anglia
EM	East Midlands Trains
GC	Grand Central
GW	First Great Western
HD	Hastings Diesels
HT	First Hull Trains
LM	London Midland
LO	London Overground
NO	Northern

SN	Southern	
SR	ScotRail	
SW	South West Trains	
TP	TransPennine Express	
VW	Virgin Trains	
XC	CrossCountry	

6.4. ALLOCATION & LOCATION CODES (* unofficial code)

Code	Location	Operator
AK	Ardwick (Manchester)	Siemens
AL	Aylesbury	Chiltern Railways
CF	Cardiff Canton	Arriva Trains Wales/Pullman Rai
CH	Chester	Alstom
CK	Corkerhill (Glasgow)	ScotRail
CZ	Central Rivers (Burton)	Bombardier Transportation
DY	Derby Etches Park	East Midlands Trains
EX	Exeter	First Great Western
HA	Haymarket (Edinburgh)	ScotRail
HT	Heaton (Newcastle)	Northern
IS	Inverness	ScotRail
MN	Machynlleth	Arriva Trains Wales
NC	Norwich Crown Point	National Express East Anglia
NH	Newton Heath (Manchester)	Northern
NL	Neville Hill (Leeds)	East Midlands Trains/Northern
NM	Nottingham Eastcroft	East Midlands Trains
OO	Old Oak Common HST	First Great Western
PM	St Philip's Marsh (Bristol)	First Great Western
RG	Reading	First Great Western
RU	Rugby Rail Plant	Colas Rail
SA	Salisbury	South West Trains
SE	St Leonards (Hastings)	St Leonards Railway Engineering
SJ*	Stourbridge Junction	Parry People Movers
SN*	MoD Shoeburyness	Ministry of Defence
SU	Selhurst (Croydon)	Southern
TM	Tyseley Locomotive Works	Birmingham Railway Museum
TS	Tyseley (Birmingham)	London Midland
WN	Willesden (London)	London Overground
XW	Crofton (Wakefield)	Bombardier Transportation
ZA	RTC Business Park (Derby)	Railway Vehicle Engineering
ZB	Doncaster Works	Wabtec Rail
ZC	Crewe Works	Bombardier Transportation
ZD	Derby Works	Bombardier Transportation
ZG	Eastleigh Works	Knights Rail Services
ZH	Springburn Works Glasgow	Railcare
ZI	Ilford Works	Bombardier Transportation
ZJ	Marcroft, Stoke	Axiom Rail
ZK	Kilmarnock Works	Brush-Barclay
ZN	Wolverton Works	Railcare
ZR	York (Holgate Works)	Network Rail